A GOURMET'S GUIDE TO

MUSHROOMS
&
TRUFFLES

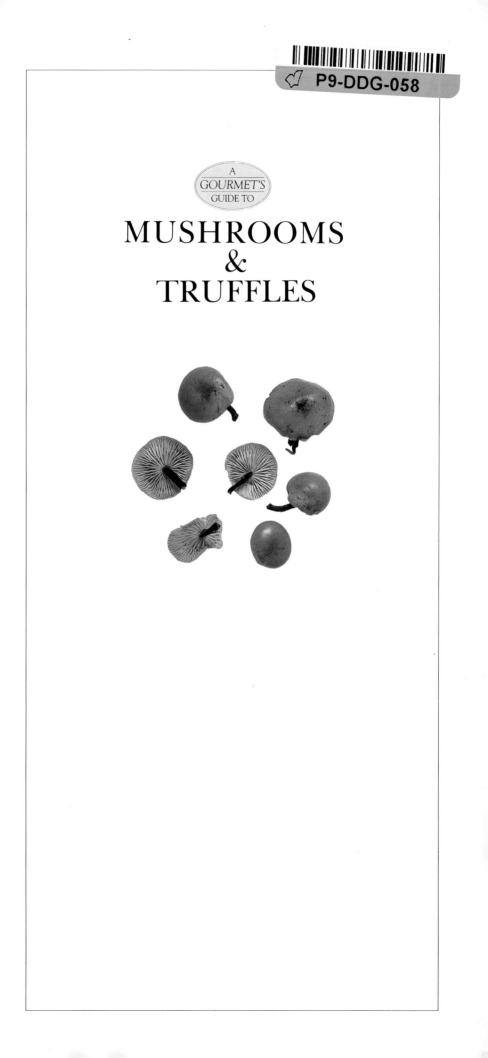

A GOURMET'S GUIDE TO

MUSHROOMS
&
TRUFFLES

JACQUI HURST
LYN RUTHERFORD

Photography by
JACQUI HURST

HPBooks
a division of
PRICE STERN SLOAN
Los Angeles

This book was created by Merehurst Limited
Ferry House, 51/57 Lacy Road, London SW15 1PR

© 1991, Salamander Books Ltd.

Published by HPBooks
A division of Price Stern Sloan, Inc.
Los Angeles, California
Printed in Belgium by Proost International Book Production, Turnhout

9 8 7 6 5 4 3 2 1

Library of Congress Cataloging-in-Publication Data

Hurst, Jacqui.
    A gourmet's guide to mushrooms & truffles / by Jacqui Hurst, Lyn Rutherford.
        P.   cm.
    ISBN 0-89586-851-2 :
    1. Cookery (Mushrooms)   2. Cookery (Truffles)   3. Mushrooms, Edible—Identification.   I. Rutherford, Lyn.   II. Title.
    III. Title: Gourmet's guide to mushrooms and truffles.
    TX804.H87   1990
    641.6'58—dc20   90-4340
                                                                    CIP

This book is printed on acid-free paper.

Commissioned and Directed by Merehurst Limited
Photography: Jacqui Hurst
Home Economist: Lyn Rutherford
Color reproduction by Kentscan, England

Contents

Introduction

Today, it's possible to buy quite a selection of wild mushrooms. In recent years, high-class produce stores and supermarkets have started selling cepes, chanterelles, horn of plenty, oyster mushrooms, hedgehog fungus and sometimes even truffles, when in season. But it's much more fun foraging for your own mushrooms—and there's a bonus in the low price (they're free)! Many people shudder at the thought of eating wild fungi—a pity, since there are delicious species to be found in woods, fields and gardens.

You will learn a little about where and when to look for fungi, what to look for, and how to go about preparing and storing your find. But *A Gourmet's Guide to Mushrooms & Truffles* is not a field guide, and it should not be your only resource. The Appendix on page 118 tells you how to find other mushroom enthusiasts in your area and recommends some field guides.

The most fundamental rule of mushroom hunting is this: *Never eat a wild mushroom unless you are quite sure of its identity.* None of the traditional tests (seeing if the fungus blackens silver, for example) can protect you from poisoning. As more and more species are named, narrowing all the possibilities down to just one may seem an overwhelming task. But because a hunter has the limited aim of recognizing a few select varieties, identification is always a choice in a context. Moreover, some varieties are quite distinctive, since there are few dangerous lookalikes. Start with these, then enlarge your repertoire as you gain experience and confidence.

If you are simply too nervous about potential misidentification to gather your own wild fungi, at least buy some wild varieties to appreciate their flavors. The selection of recipes on pages 64-117 offers original ideas for delicious soups, starters, main courses, salads and accompaniments, using both cultivated and wild varieties of mushrooms.

When and Where They Grow

Mushrooms are magical. They have an uncanny habit of springing up, maturing and vanishing within the space of a few days or a couple of weeks. The cepe has a growing cycle of about 2 weeks, whereas other kinds appear and disappear within 24 hours. Some types favor the same spot year after year; others do not. The unpredictable nature of mushrooms adds to the excitement of the hunt, and to the pleasure of gathering them successfully. But you can improve the odds of a satisfying find by training yourself to see the regularities that lie behind the apparent capriciousness.

One basic principle is that a mushroom is likely to reappear at some time, since its hidden vegetative state persists in the soil or tree where the visible part catches your eye. (This rule does not apply if the mushroom is part of a changing habitat, as in a forest recovering after a fire or on a decaying log which eventually must disintegrate.) Some mushrooms are parasites; others live on dead material, while still others form symbiotic relationships with specific plants— *Suillus grevillei* (see page 40), for example, will grow only in association with a particular tree. And each species has its preferred time to fruit. By learning to recognize their plant allies

Cultivated varieties

Oyster mushrooms

White or Button mushrooms

Brown mushrooms

Yellow pleurotes
(Pleurotus ostreatus)

and victims, their favored environments and seasons, you can hone in on the mushrooms.

Mushroom enthusiasts are rather like farmers, always moaning about the weather: "It's not a very good season—it's too dry (too warm, too cold)," or "There was a hard frost last night that will stop them growing." But despite such grumbles and dire predictions, you can usually find some fungi. In any area, the climate will determine the particular collection you may discover; and every year, there are certain triggers of rain or temperature that evoke quite a variable response. In an unusual year, some species may be rare or absent. In general, fungi favor a combination of warmth and damp; a long, dry summer followed by a wet, mild fall encourages all kinds of mushrooms to burst from the soil in profusion, whereas after a cold summer and a dry fall, they tend to be thin

on the ground.

For an area as vast and climatically varied as North America, it's hard to go too far beyond these general statements. Mushrooms do tend to avoid extremes of cold and desiccation, so major fruitings occur in spring and especially fall. On the whole, the fruiting season will be longer in the Southeast, reflecting the milder weather there, and may even extend throughout the winter in the South. In California, there's little to be found in the dry summer, but a major fruiting appears dramatically after the first rains. In Colorado and New Mexico, July and August are typically the best times.

Even today, our knowledge about the distribution of species in North America is very patchy; there is no simple way to describe where and when you will find a particular mushroom. A field guide can help explain regional differences.

A few other points to keep in mind: When mycologists (mushroom experts) refer to *wood mushrooms,* they don't mean you have to penetrate into the depths of some dark, gloomy forest to find your quarry. These mushrooms need a certain amount of light and warmth, and tend to favor the perimeters of woods and the clearings within wooded areas. *Meadow mushrooms,* on the other hand, grow on established pastures which have been grazed by horses, cows or sheep, because they rely on well-manured soil for their nutrients.

Collecting
People often think that you have to get up at the crack of dawn to collect mushrooms, but in fact any time of day will do. The only advantage of going mushrooming first thing in the morning is that you can gather meadow mushrooms, russulas or cepes before the insects have had their share.

If you're a beginner, it's a good idea to go collecting with an experienced companion or join an organized foray run by a local group. You'll learn a lot and have the peace of mind that comes from knowing that your collection has been verified by an expert. It's also important to buy a reliable field guide with good illustrations, so you can become familiar with all types of fungi. Learning to recognize the toxic types is crucial, since you obviously can't afford to confuse these with edible kinds. Identification is always a choice in the context of what else may occur.

The Appendix on page 118 gives advice on finding a local society and on choosing a field guide.

When you select an area for your collecting, remember to respect private property and to acquaint yourself with local regulations. Some parks protect all forms of wildlife, including mushrooms, while others may limit the amount you may pick or may require a permit.

Take a flat-bottomed basket with a selection of small containers to keep the different species apart. Waxed paper bags, their ends closed with a twist, preserve specimens well. Don't put the fungi in your pockets or in plastic bags; they can easily be damaged and are liable to sweat and rot.

rotting.

It is preferable to gather mushrooms in dry weather. Don't pick old, maggoty specimens or very young ones which have not yet developed their distinguishing characteristics. Never cut through the stalk; for full identification, you must get the whole stem from beneath the soil and see if the mushroom's base is tapering, bulbous, or perhaps encased by a bag-like sheath known as a *volva.*

The best way to pick a fungus is to twist it gently until it breaks free. Sometimes you will need a knife to loosen the surrounding soil, but take care not to damage the *mycelium*—the fine, threadlike root system.

Note the cap shape, size, color and texture. Check to see if the underside has gills similar to those of the meadow mushroom (see page 12), a spongy mass of tubes and pores characteristic of the boletes (see pages 36-41), or spines like the hedgehog fungus (see page 43). If the mushroom has gills, note their color and see if they're *decurrent*—descending down the stem. Look at the margin (the edge of the cap): is it smooth, flared, wavy or in-rolled? Does the cap have a raised bump (an *umbo*) in the center?

Study the stem carefully, too. Does it have a ring around it? If so, is the ring floppy, soft or even double?

Check to see if the fungus bruises when touched; break the flesh to see if it exudes a milky juice like the saffron milk cap (see page 26).

A mushroom's smell is often difficult to ascertain, though some types have quite a distinctive odor. The horse mushroom (see page 13), for example, usually smells of anise; *Russula xerampelina* (see page 33) has a distinct crablike aroma. The rest have a faint scent which is easily masked.

When you gather the tree species, generally known as *bracket fungi,* pry them off the wood gently with a knife. Then, as with other mushrooms, check all the salient features.

Before slicing up your mushrooms and adding them to the skillet, double-check your identification. Be alert to the possibility that adjacent mushrooms may be different. *Be 100 percent certain that you know exactly what you have collected, because mushroom poisoning is extremely painful and may be*

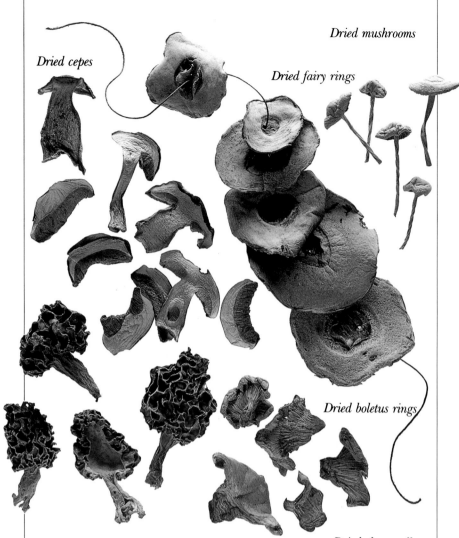

Dried cepes

Dried mushrooms

Dried fairy rings

Dried boletus rings

Dried morels

Dried chanterelles

fatal. Even when you are in no doubt, it's good practice to eat only a small quantity of a new kind of mushroom at first, since you may be allergic to it. Most ill effects will be felt quite quickly—within a couple of hours—but reaction to an amanita (see page 42) may be delayed up to 8 hours.

Preparation

It is considered a crime to wash mushrooms, but some species—such as morels (see page 44) and cauliflower fungus (see page 50)—require it, since soil, pine needles and leaves get lodged in the holes, and a host of small insects can lurk in the folds and crevices. In general, though, the best way to clean mushrooms is to wipe them with a damp cloth to remove any dirt. Separate the stem from the cap to check for invading insects; the stalks tend to become maggot-ridden first. Do not peel mushrooms—the skin contains a lot of the nutritional goodness and flavor.

To prevent raw mushrooms from discoloring, cut and coat them with lemon juice.

Storing

Handle mushrooms as little as possible to avoid bruising them. Store in a brown paper or waxed paper bag in the crisper drawer of the refrigerator. Use the mushrooms within a couple of days, since they do not keep well.

Preserving

When a good season brings a glut of mushrooms, you're bound to pick more than you can eat. Drying the surplus is the best way to preserve the flavor for a long period.

For drying, use only very fresh mushrooms. Remove any blemishes and clean thoroughly to remove all dirt. Cut large specimens (a big bolete, for example) into thin slices; leave small ones, such as fairy rings, whole.

If the weather is warm and bright, you'll be able to sun-dry fungi. But if you live in a cool, damp climate, you'll have to dry the mushrooms on a wire rack in a warm oven. Let the mushrooms dry until they shrink and become brittle (usually a couple of days), then pack them in an airtight jar.

Dry mushrooms in a low oven with the door slightly ajar, but if the temperature exceeds 140F (60C), they will cook, blacken and lose their flavor. A third method is to thread the mushrooms on fine string and hang them in a warm place to dry; morels take especially well to this treatment.

To reconstitute dried mushrooms, soak in warm water 20 to 30 minutes.

If you have dried some of the more strongly-flavored mushrooms, it is better to grind them to a powder and use this as a seasoning; the concentrated flavor is wonderful in soups and vegetable dishes.

You can also preserve mushrooms by pickling them in vinegar or making them into ketchup.

Freezing
Though you can freeze raw mushrooms, it diminishes their flavor, and they only keep for a month. The best way to freeze mushrooms is to make *duxelles*. Sauté some finely chopped onion and pressed garlic in butter until softened. Thinly slice or chop the mushrooms, add to the saucepan and cook until the liquid has evaporated. Season with salt, pepper, nutmeg and a handful of finely chopped parsley. Let cool, then transfer to a freezer container, seal and freeze. Use frozen as part of the filling for a chicken pie, as a flavoring for soups, stews and casseroles, or as a base for sauces.

If you do want to freeze whole raw mushrooms, use only fresh, firm specimens. Clean them and freeze in a single layer, uncovered, until solid; then pack in plastic bags or freezer containers and return to the freezer.

Cooking
Mushrooms are virtually calorie-free. They are delicious raw—especially if marinated in olive oil and herbs—and are lovely simply grilled.

Meat, game, fish, soups and sauces can all be enhanced with the addition of mushrooms. And of course, mushrooms are delicious on their own—try them stuffed, baked or made into tempura. All mushrooms, especially truffles, have an affinity for eggs; cepes, chanterelles and blewits are excellent cooked with potatoes. All types benefit from a squeeze of lemon juice during sautéing—it helps to bring out the flavor.

When substituting wild species for their cultivated relatives, you'll need a greater quantity, since wild types have a higher moisture content and shrink more. Shaggy manes, chanterelles, parasol mushrooms and hedgehog fungi in particular, contain an enormous amount of water; when sautéing these types in butter, beware of spattering juices. If there's an excessive amount of liquid, drain it off and reserve it for a stock.

Wild mushrooms' extra flavor compensates for their lack of size, especially where dried fungi are concerned. It can look very stingy when you soak only 2 or 3 shriveled slices, but their richness is sufficient to permeate a whole dish.

Using This Book
Nearly all the mushrooms described on the following pages grow wild in North America. Many have distinctive features which make them easy to identify. Where possible, I have avoided repetition by noting the common characteristics shared by a particular family of mushrooms.

Few kinds of fungi are truly lethal, but a much larger number are toxic to some degree or to some people in some circumstances. The mushrooms in the genus *Amanita* are in category all by themselves, though there are other deadly mushrooms besides these and some amanitas are in fact choice edibles. Nonetheless, these are mushrooms every collector should be able to recognize and beginners should avoid. The deadly ones are beautiful, conspicuous and, in some parts of the country, abundant. They are sometimes mixed in with *Agaricus*, and they masquerade as puffballs when young. Amanitas are briefly described on page 42; for details and illustrations, consult a field guide (see Appendix).

1/2 life size

(Agaricus bisporus)

Meadow mushroom
(Agaricus campestris)

Meadow Mushroom

The Meadow Mushroom *(Agaricus campestris)* is the best-known wild mushroom, though the good old days when they carpeted the meadows are gone. They are still common in pastures, but tend to favor dense clumps of grass and only become visible when you are almost on top of them. You can find them from midsummer through fall or, in California, from fall through winter.

The meadow mushroom's cap is 1-1/2 to 4 inches across, silky white turning to cream-colored. The pale pink gills turn reddish, then brown when the button eventually opens; take care to avoid confusion with a white-gilled deadly amanita which can be found in some of the same environments. The meadow mushroom's stem tapers towards the base and has a delicate ring; taste and smell are pleasantly mushroomy.

Collecting and Storing

To collect meadow mushrooms, you need to train your eyes to scan the ground a yard or so in front of you;

nestled in the grass, they are easy to miss or walk on. Check to make sure there are no warts on the caps. If the mushrooms have pure white gills or stain yellow at the stem base, throw them away.

Eat these mushrooms fresh.

Preparation and Cooking

Do not peel these mushrooms or their flavor will be slightly diminished; just wipe with a damp cloth and cut off the dirty base.

The best way to eat meadow mushrooms is to fry them in bacon fat—perhaps with a little garlic—as soon as you return home. For a decent meal, you will need a panful; because of their high moisture content, the mushrooms shrink to a third of their original volume.

Eat them on their own; add bread crumbs and spices and use the mixture to fill filo pastry packets; or thicken the juices with egg yolks and cream to make a sauce. Meadow mushrooms are delicious served with shellfish, especially scallops.

Agaricus bisporus

Parent of most cultivated varieties of the traditional store-bought mushroom (see page 58), this mushroom can also be found growing in vacant lots and on compost heaps during the

fall. It is similar to the meadow mushroom, but the 2- to 4-inch-wide cap is a dirty buff color rather than white. *Agaricus bisporus* is good to eat; cook the same as meadow mushrooms.

Horse Mushroom

A giant, meaty fungus recognizable by its distinct anise smell, the Horse Mushroom *(Agaricus arvensis)* is a large version of the meadow mushroom. It differs from meadow mushrooms in that its creamy-white cap and stem discolor and bruise yellow.

Great care must be taken not to confuse this mushroom with the toxic Yellow Stainer *(Agaricus xanthodermus)*, which causes violent illness if eaten. To differentiate between the two, dig up the whole stem from beneath the ground, then cut through the base; if it turns bright yellow, you have the poisonous mushroom. If you have *any* doubts about your identification, throw the mushroom away.

The horse mushroom can grow in vast quantities, often in rings, in old meadows or groves. It may reappear year after year in the same locality, from midsummer through mid-fall or, in California, from fall through winter. The cap, dome-shaped when young, can eventually expand up to 10 inches across. The gills are white at first, turning to pinkish-gray and then to chocolate-brown. The stem has a large, floppy double ring; the flavor is strongly mushroomy.

Collecting and Storing

Take extra care picking this species. Check the base to see if it turns yellow when cut (see above) and sniff the cap for an anise scent, which is usually present. If you are confident you have found the horse mushroom, take it home and add it to the cooking pot. For maximum flavor, eat these mushrooms the day they are picked.

Preparation and Cooking

There is no need to peel horse mushrooms—just wipe away the dirt with a damp cloth. You can use the stems, but they tend to be maggot-ridden and become fibrous with age.

A single large, flat cap makes a hearty meal, especially at breakfast. Horse mushrooms are wonderful simply coated in herb butter and grilled; their huge size also makes them ideal "nests" for stuffed quail—an unusual dish to serve at a dinner party.

If the horse mushrooms you find are still dome-shaped, stuff them with a savory filling and bake them.

Horse mushrooms can be used in most recipes, but they're particularly good for croustades (see page 75).

Horse mushroom (Agaricus arvensis)

1/2 life size

Agaricus silvicola

This woodland species is similar in appearance to its big brother, the horse mushroom: it discolors and bruises yellow and has a distinct smell of anise. *Agaricus silvicola* grows in coniferous and deciduous woods. It is found from midsummer through late fall or, in California, from fall to winter.

The dry, creamy-white cap is 2 to 4 inches across. The pinkish-gray gills darken to chocolate-brown with age. The silky white stem has a button base and a floppy ring.

Great care must be taken not to confuse *Agaricus silvicola* with lethal *Amanitas* (see page 42) and the toxic Yellow Stainer *(Agaricus xanthodermus)*.

Collecting and Storing
When gathering this mushroom, cut a couple of the young buttons in half to see if the flesh is yellow at the base; also check to see if any of them have white gills. If either of these features is present, you have picked poisonous mushrooms. If possible, get your identification verified by an expert before cooking. Some individual gastric reactions have been reported, so sample only a small quantity at first.

Agaricus silvicola should be eaten fresh, as its flavor deteriorates quickly after picking.

Preparation and Cooking
Prepare and cook this tasty woodland species in the same way as meadow or horse mushrooms (see pages 12-13). The young caps are lovely coated in batter and deep-fried.

Agaricus haemorrhoidarius

This mushroom is very similar in appearance to *Agaricus silvaticus* (below), but is found in deciduous woods. The cap is covered in downy reddish-brown scales; the flesh turns red when cut. The stem is slightly bulbous, with a broad ring attached. *Agaricus haemorrhoidarius* appears in late summer and fall or, in California, in fall and winter.

Collecting and Storing
Pick fresh young specimens. Like all edible members of *Agaricus*, this species should be eaten fresh; if you wish to preserve the mushrooms, freeze them as duxelles (see page 11). Observe the same caution as for *Agaricus silvicola*, since gastric upsets occur in some people.

Preparation and Cooking
Cut off most of the stalk and clean the cap with a damp cloth. Fry with shallots, and serve on toast.

Agaricus silvaticus

The mature caps of this species look very much like those of the meadow mushroom, though *Agaricus silvaticus* prefers the shade. You can find it growing in coniferous woods throughout the fall.

The white cap, covered with tiny reddish-brown speckles, measures 2 to 4 inches across. The gills are pinkish-gray, turning blackish with age. The rather scaly, bulbous stem eventually becomes hollow, and has a floppy ring. The young buttons redden when they are cut, but the older specimens, with brown flesh, do not discolor. Don't let this reddening put you off; *Agaricus silvaticus* has an excellent flavor and a pleasant aroma.

Collecting and Storing
Only collect fresh young buttons; the larger ones tend to be infested with insects. They should be eaten fresh, but if you have a surfeit of them, freeze them as duxelles (see page 11). Sample these mushrooms in very small amounts the first time you eat them.

Preparation and Cooking·
Use small buttons whole, but discard the tough, fibrous stalks from older specimens. Remove any dirt with a damp cloth.

These mushrooms are suitable to use in any mushroom dish.

Agaricus silvaticus

Agaricus haemorrhoidarius

Agaricus silvicola

1/2 life size

Prince

The Prince, *Agaricus augustus*, is a large mushroom (up to 12 inches across) with a delightful almond odor when young. The cap is typically golden-colored and covered in feathery, brown-tipped scales. The prince has a large membranous ring; the stem below is often scaly, bruising yellowish. You may find this mushroom in grassy areas and woods and on roadsides in summer and early fall (into mild winters in California). It is one of the meatiest and most esteemed of all mushrooms and may be prepared like the horse mushroom (see page 13).

Honey Mushroom

Honey Mushroom *(Armillaria mellea)* is extremely common and can be found from midsummer through late fall, growing in large tufts on living and dying trees, old stumps, buried branches or dead roots. It is very destructive in parks and gardens, responsible for killing all kinds of trees by causing white rot.

This mushroom is variable in appearance; its cap can be honey-colored and covered with dark scales, rusty-red or dark brown. Whatever its color, the cap is usually darker in the center. Cap size varies from 1-1/4 to 6 inches across. The gills are creamy yellow. The stem is white to yellow, sometimes spotted with rusty flecks, and has a large, soft ring and sometimes a button base. The stems often fuse together in clumps.

Honey mushroom is also known as Bootlace Fungus, because it spreads by means of long black cords (called *rhizomorphs*); you'll always find these under the bark of infested trees.

A similar species, *Armillaria tabescens*, which also grows in clusters on old stumps, can be distinguished from honey mushroom by its ringless stem. It is edible, but must be cooked.

Collecting and Storing

Pick very young clusters. You can find this fungus in large quantities, but it does not keep well and is not worth drying. Don't discard uncooked scraps in your garden; you'll be giving an unwelcome guest a foothold.

Preparation and Cooking

Honey mushrooms should *never* be eaten raw; also reject large, old specimens, because they can be very toxic and cause serious stomach upsets. Eat only the caps.

Cooked honey mushrooms become rather soggy and slippery, since they absorb all the frying fat or oil. They aren't one of the best edible species, but for some there is a certain vindictive pleasure to be gained from eating a fungus which destroys many of our beautiful trees!

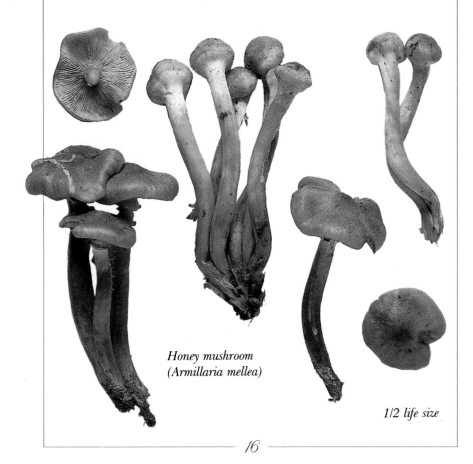

Honey mushroom
(Armillaria mellea)

1/2 life size

Fairy ring mushroom (Marasmius oreades)

Fairy Ring Mushroom

One of the most common grassland species, the Fairy Ring Mushroom *(Marasmius oreades)* is found growing in large numbers in pastures and on lawns. As the name suggests, it often forms a ring. In fact, this fungus is much disliked by gardeners precisely because it produces "fairy rings"— bare circles bordered by dark grass— at the spots where it grew.

Fairy ring is a well-known and reliable mushroom, but great care should be taken to avoid misidentification and confusion, especially with the poisonous white lookalike *Clitocybe dealbata*. Other mushrooms form rings, so this characteristic alone is not enough for a positive identification of fairy ring.

Fairy ring mushrooms grow on slender, straw-colored stems which are tough and leathery. The wide, bell-shaped cap expands as it matures and ranges in size from 3/4 to 2 inches across, but always retains a small center umbo. Moist, younger mushrooms vary in color from rusty-brown to tan; as they dry, they fade to pale buff with a fawn center. The smell is slightly haylike.

Fairy ring can appear in late spring and continue growing almost until winter. In California, these mushrooms can be found year round.

Collecting and Storing
Gather only fresh young mushrooms; pick the whole plant to ensure that you have the correct species. Fairy rings are ideal for drying and seem to retain their flavor longer than other mushrooms. They are perfect to store for the winter, when they can be used to flavor soups, sauces and casseroles.

Preparation and Cooking
Discard the tough, fibrous stem and wash the cap as little as possible. A basic method of cooking these delicious mushrooms is to fry them gently in butter with salt, pepper and a squeeze of lemon juice, then let them simmer 15 minutes. They can then be eaten on their own or added to an omelet.

Fairy ring mushrooms can be made into cream of mushroom soup. They also add an interesting flavor to rice or ground beef fillings for peppers.

Chanterelle (Cantharellus cibarius)

Chanterelle

The Chanterelle *(Cantharellus cibarius)*, extensively collected in North America (especially on the West Coast), is becoming increasingly common in produce shops and even some supermarkets. Large quantities are exported to Europe, where this mushroom is highly prized. In France (where it's also called *girolle*), the chanterelle is commonly sold in markets and grocery stores and served in restaurants.

Funnel-shaped mushrooms with stems that taper toward the base, chanterelles grow in all kinds of woods, though they prefer stands of conifers and oak. You can find them in clusters from midsummer until the beginning of winter.

The cap measures from 1-1/4 to 6 inches across; it is depressed in the center and has a curly edge. Instead of gills, the chanterelle has forked ridges—like veins—running halfway down the stem. As the mushroom ages and dries, its beautiful bright egg-yolk yellow color fades to a paler yellow. Its scent is reminiscent of apricots; the taste is slightly aromatic.

Take care not to confuse this fleshy mushroom with the rather unpalatable False Chanterelle *(Hygophoropsis aurantiaca)*, which is fragile in comparison and has ordinary gills and no smell. Also avoid confusion with the poisonous Jack O'Lantern *(Omphalotus)* species, which usually occurs in clusters and has a dirty yellow color and well-developed gills.

Collecting and Storing

Chanterelles have a habit of hiding under leaf litter in the woods, but you soon learn to recognize their characteristic humps. When you find one, there are usually more close by. Discard the muddy bases to keep the mushrooms in the basket from getting covered with grit.

If you do not have time to go mushroom hunting, you can buy chanterelles at some markets; the price varies widely, depending on availability.

Chanterelles will stay fresh up to a week if kept in the refrigerator. They can be dried, but do not retain their flavor as well.

Preparation and Cooking

Wipe the caps with a damp cloth and brush the dirt out of the ridges. These mushrooms are seldom attacked by insects, so it is only necessary to slice the larger ones lengthwise into strips. Chanterelles release a lot of liquid while cooking and shrink to about a third of their original volume.

Chanterelles go well with eggs, bacon and potatoes, and they are wonderful in sauces. A simple way to prepare them is to fry them slowly in butter or olive oil, with garlic and salt.

Yellow Leg

The Yellow Leg *(Cantharellus infundibuliformis)* is fairly common in coniferous woods and damp areas. The clustered dark brown caps are rather difficult to see, but if you find one mushroom, it's worth ferreting around in the surrounding leaf litter, since yellow legs often grow in large groups. They first appear in midsummer and, in a good year, continue to grow until the beginning of winter.

The cap is slightly convex when young, becoming funnel-shaped with a wavy margin as it develops; it measures 1-1/2 to 2 inches across. In place of gills, there are narrow, branching veins—yellowish at first, then gray—descending down the compressed, dirty yellow stem. The flesh is thin. These mushrooms smell faintly aromatic but taste bitter and are unsavory if eaten raw.

Collecting and Storing
If you are lucky, you should be able to gather quite a number of yellow legs; if you can only find a few, mix them with other kinds. Before putting these mushrooms in your basket, cut off the earthy bases.

If refrigerated, yellow legs will stay fresh up to a week. They are easy to dry; store them to use later as a flavoring for stews.

Preparation and Cooking
Wipe the caps with a damp cloth and brush away any grit embedded in the veins (or wash, if necessary). Because yellow legs are quite small, they can be used whole, though the stalks of older specimens are leathery and rather chewy and should therefore be removed.

Yellow legs need long, slow cooking. Use them as you would chanterelles (see opposite); make into a sauce or add to egg dishes. They look attractive in clear soups and are tasty stir-fried with broccoli.

Yellow Leg (Cantharellus infundibuliformis)

Horn of Plenty

Horn of Plenty *(Craterellus cornuco-pioides)*, or Black Trumpet as it is sometimes called, is considered a great delicacy; in France, it is described as *la viande des pauvres* ("poor people's meat").

This fragile, trumpet-shaped mushroom has a waxy, charcoal-gray outer surface, while the inside is a velvety blackish-brown. These somber colors may explain the French common name *trompette de la mort* ("death trumpet"). Another characteristic feature is the lack of gills. The cap has a flared, wavy margin and is 3/4 to 3 inches across.

Horn of plenty can be found growing in large groups in damp deciduous woods, especially under oak or beech, during summer and fall. On the West Coast, it fruits in winter and spring under hardwoods or, in northerly regions, under conifers.

Collecting and Storing
Horn of plenty's ashen color makes it very difficult to spot, but once you've discovered one mushroom it's worth searching through the leaf litter to find a good collection. Horn of plenty will stay fresh in the refrigerator 4 to 5 days, but can be dried very easily and retains its flavor for later use in soups and stews. The dried mushrooms can also be powdered and used as a seasoning for soufflés or terrines.

If you can't find this mushroom growing wild, you may be able to locate it in produce markets, at least on the West Coast. It is also dried.

Preparation and Cooking
Cut down one side so you can open up the funnel and remove any dirt trapped in the hollow stem with a brush or damp cloth.

Horn of plenty is delicious simply sautéed in butter with a squeeze of lemon juice; add a little cream and serve on toast for breakfast. These mushrooms are also wonderful stir-fried with strips of bell pepper and halibut, cod or salmon.

Craterellus cinereus

These funnel-shaped mushrooms are relatively uncommon. They bear a slight resemblance to horn of plenty (above), but are smaller, brownish-gray in color rather than black, and have a compressed stem. They grow in clusters under broad-leaved trees and are difficult to find among the leaf litter.

This is a good, reliable species; if you are able to find some, cook them as for horn of plenty.

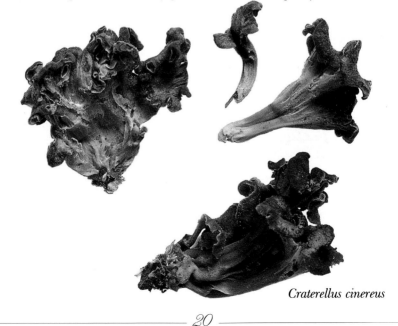

Craterellus cinereus

Horn of plenty (Craterellus cornucopioides)

Sweetbread Mushroom

Sweetbread Mushroom *(Clitopilus prunulus)* is a very common mushroom, found growing in woodland glades and on shrubby meadows from midsummer until the end of fall. Its strong, mealy smell and taste are tempered by cooking.

The sweetbread mushroom's 1-1/4- to 4-inch-wide cap is irregular in shape, convex to depressed, often wavy-edged; it has a matte, suedelike surface which turns from white or pale cream to grayish with age. The white gills descend down the stem and become pink as the mushroom matures.

Though this is a good edible mushroom, great care must be taken not to confuse it with three poisonous species which closely resemble it in both appearance and smell: *Clitocybe dealbata, Clitocybe rivulosa* and the rarer *Entoloma sinuatum.*

Collecting and Storing
When you collect the sweetbread mushroom, pick the whole mushroom and use a field guide to help avoid confusion with the poisonous species named above. If possible, have an expert check your identification; if there is any doubt, leave the questionable mushroom behind. Sweetbread mushrooms will stay fresh for a couple of days in the refrigerator.

Preparation and Cooking
Use only fresh young specimens. Cut away the earthy bases and wipe the caps with a damp cloth to remove any dirt.

Slice the mushrooms and fry in butter with a finely chopped onion and a handful of fresh herbs, such as marjoram, chives or thyme. When most of the juices have evaporated, add a spoonful of yogurt. (Or add some garlic and use sour cream instead of yogurt.) Serve as a starter or on toast as a light snack.

Alternatively, cook the mushrooms in olive oil with onions, garlic and tomatoes; season with pepper and fresh basil and serve with spaghetti.

Some people find this mushroom's strong flavor rather overpowering; if you belong to this group, mix your sweetbread mushrooms with cultivated types.

Velvet Foot

As the common name implies, Velvet Foot *(Flammulina velutipes)* has a dark, velvety stem. This species is one of the few mushrooms which can survive frost; even if frozen solid, it revives upon thawing. It grows from late fall through spring in the East; because it appears so late, there's little chance of confusing it with other fungi.

Velvet foot is a rather fragile, fairly common mushroom which grows in tufts on decaying stumps and dead branches, especially those from elm trees. If you have a fireplace, you're just as likely to find it in your log pile as anywhere else. The sticky, shiny cap is orange and honey-yellow, darkening towards the center, and ranges from 1/2 to 2 inches across. The gills are pale yellow. The stem is tan at the top, blackish-brown at the base, and covered in dense, velvety hairs.

Velvet feet smell and taste pleasant, and often occur when few other species are available.

Collecting and Storing
Pry the clusters from the wood with a knife. The only way to preserve these mushrooms is to dry them and store them airtight until needed.

Preparation and Cooking
Use only the tiny caps, first being sure to wash off the sticky coating. These mushrooms can be sautéed in butter or added to casseroles, but they have a rather rubbery texture. It's preferable to dry the caps and grind them to a powder, then use them as a flavoring for soups and sauces (add just before serving).

Sweetbread mushroom (Clitopilus prunulus)

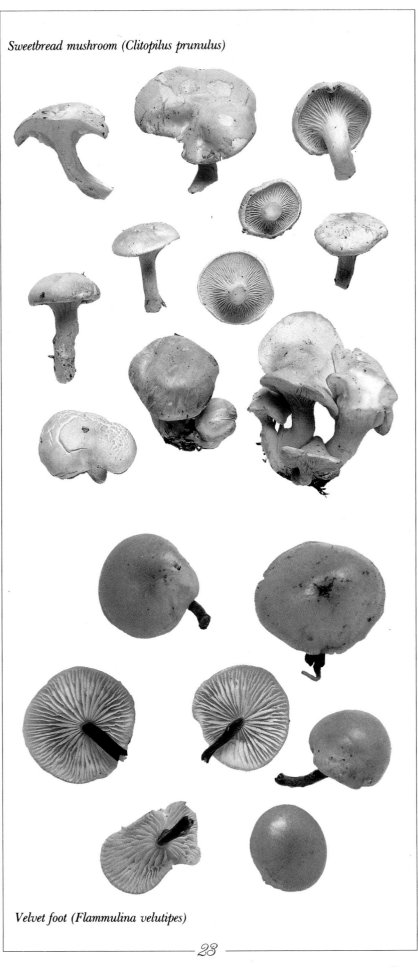

Velvet foot (Flammulina velutipes)

Shaggy Mane

Common in fields, along grass tracks, and on compost heaps and recently disturbed ground, the Shaggy Mane *(Coprinus comatus)* can be found in vast quantities. It is more typically seen, however, growing in small tufts on lawns or roadsides during summer and fall.

This mushroom is also called Shaggy Ink Cap, Inky Top and Lawyer's Wig; the last name derives from the fact that the cylindrical white cap, 2 to 6 inches tall, is covered with feathery curls like those on a British jurist's wig. As the cap matures, it becomes limp; the white gills turn pink, then black as they begin to dissolve, leaving a dripping button on top of a long, smooth white stem (the stem has a tiny white ring when young). Shaggy manes have a mild, pleasant flavor.

Collecting and Storing
Gather only young mushrooms, with closed caps and white gills, for eating. Shaggy manes are extremely fragile and must be cooked as quickly as possible after picking, since they start to disintegrate within a couple of hours.

Preparation and Cooking
To prepare shaggy manes, discard the hollow stem. Wipe the caps carefully to remove any dirt, but do not wash them as they may collapse.

To make a thin, delicate soup, add the white caps to chicken or vegetable stock; season with salt and pepper, enrich with cream, and sprinkle with finely chopped parsley just before serving.

Shaggy manes are delicious baked with a chunk of butter, some finely chopped shallots, salt, pepper and a hint of garlic; eat them piping hot, with homemade bread to mop up the juices.

Alternatively, cut the caps lengthwise into strips and lightly cook in butter. Serve on toast, or place the slivers and their juices on top of eggs in individual buttered ramekins, set in a hot-water bath and bake in a 325F (165C) oven just until the eggs are set.

Shaggy manes are excellent with fish, especially turbot, sole and whiting. They also can be battered and fried to be served with fried fish.

Inky Cap

It is the Inky Cap *(Coprinus atramentarius)* that gave the ink cap group of mushrooms its name: at one time, ink was made by boiling the blackened caps with water and cloves. Inky cap is distinguished from its cousin, the shaggy mane, by its shorter (1-1/4- to 2-3/4-inch-tall), smooth, dirty gray cap. The gills are white, turning black with age.

Inky cap grows in clusters near the bases of trees in fields and parks from spring through late fall. The young buttons with white gills are delicious, *but they must not be eaten with alcohol:* the combination will cause palpitations and vomiting. This reaction can occur even when mushrooms and alcohol are consumed as much as 2 days apart. *Plan on an alcohol-free week when you eat these mushrooms.*

Mica Cap

The Mica Cap *(Coprinus micaceus)* has a short, tawny-brown cap, 1/2 to 1-3/4 inches tall; when young, it's covered with a fine, powdery dust. Mica cap is considered edible but hardly worth collecting.

Mica cap (Coprinus micaceus)

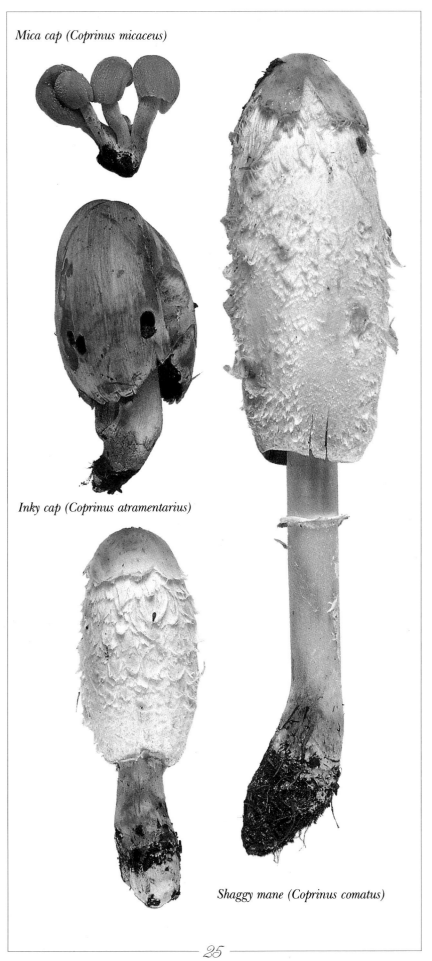

Inky cap (Coprinus atramentarius)

Shaggy mane (Coprinus comatus)

Saffron Milk Cap

The Saffron Milk Cap *(Lactarius deliciosus)* is much sought after in Europe, especially in France, where it is collected in vast numbers and canned. In North America, it is found under conifers, often in abundance, from late summer through fall or, in California, in late fall and winter.

This is a thick, fleshy mushroom, bright orange when young, turning green-spotted with age. The cap measures 1-1/4 to 4 inches across and is convex or slightly funnel-shaped. The vivid gills are decurrent (running down) into a stout salmon-colored stem covered with dark orange blotches.

All fresh specimens of the genus *Lactarius* exude a milky liquid when the stem is cut or the gills are crushed; the saffron milk cap weeps a sweet orange milk with a bitter aftertaste. This mushroom is not quite as delicious as its specific name would lead you to assume; it's mild and slightly bitter.

A number of other edible members of the genus *Lactarius* have a following among mushroom enthusiasts. *Lactarius indigo,* which has dark blue milky-sap, is found in summer and fall throughout the East and South among oak and conifers. *Lactarius hygrophoroides,* with white milky-sap, is found about the same time in deciduous woods. The milky sap is quite distinctive for the genus, but because some members of *Lactarius* are poisonous and others are unpleasant, you should consult a field guide whenever you find these mushrooms.

Collecting and Storing
Pick only young specimens, as the older ones tend to be infested with insects. Carefully wipe (or wash, if necessary) to remove surface dirt.

This mushroom will keep several days in the refrigerator.

Preparation and Cooking
A lovely way to cook these mushrooms is to slice them and fry in olive oil with a garlic clove until light brown, then add some white wine, a mixture of fresh herbs and lemon juice and cook until the liquid is reduced to a good thick sauce. Eat immediately with fresh crusty bread.

Alternatively, add the cooked mushrooms to lamb stew; or fold them into a Madeira cream sauce and serve with pork tenderloin.

Candy Caps

A number of small, red-brown species of *Lactarius* with white or watery-white milky-sap are known as Candy Caps. One is *Lactarius camphoratus;* it is rather unpalatable to eat fresh, but if dried and powdered has a strong, spicy, curry flavor and aroma and makes a good seasoning.

Lactarius camphoratus is usually found in groups under pine trees, but sometimes you can find it in deciduous woods during late summer and fall. Its matte, chestnut-brown, 1- to 2-inch-wide cap is slightly depressed, but often has a small hump in the center. The gills are decurrent and closely spaced. The milk is quite watery and has a mild flavor. *Lactarius fragilis* is similar to *Lactarius camphoratus,* but apparently restricted to the West Coast and Southeast. In the West, it can be found from late fall through spring in the woods.

Saffron milk cap (Lactarius deliciosus)

Parasol Mushroom

At the height of summer, you can often pick your first Parasol Mushrooms *(Lepiota procera)*—a welcome find that usually heralds the beginning of the mushroom season. Parasols can grow in the same spot year after year, and in a good season there will be several flushes for you to gather.

Tall and distinctive, the parasol is one of the best edible species, with a delicious nutty flavor. From midsummer until mid-fall, it is found in woodland glades, on parkland and along the edges of trails. Its preference for open spaces makes it easy to spot. The young specimens look like drumsticks, but as they mature, the cap expands like an umbrella and may reach 16 inches across. The cap is pale buff in color, covered with feathery dark brown scales, and has a central boss; the brittle gills are cream-white. The hollow, slender stem grows from a bulbous base; it is covered with a snakeskin-like pattern and has a loose ring.

Chlorophyllum molybdites is a poisonous lookalike, best distinguished from the edible lepiotas by its greenish spore print (the spores of *Lepiota* are white). Consult a field guide for instructions on taking a spore print and for information on other distinguishing characteristics.

Collecting and Storing
Ideally, you should only pick parasols when the cap is just beginning to open, but the large parasols look so wonderful that it's difficult to resist them. Don't gather any that have started to dry out; they'll be too leathery to eat.

Parasols are best eaten the day you collect them, but they will keep in the refrigerator a couple of days. They can be dried, but tend to lose some of their flavor.

Preparation and Cooking
Discard the fibrous stalks. Carefully wipe the caps with a damp cloth; wash only if necessary.

The cup-shaped caps are ideal for stuffing and baking. My favorite way to prepare parasols, is to coat them with an anchovy and parsley butter and grill them.

Shaggy Parasol

The strongly aromatic Shaggy Parasol *(Lepiota rhacodes)* is short and stocky in comparison to its elegant big brother—but it tastes just as delicious, though the fact that the cut flesh turns red tends to make some people rather squeamish.

The shaggy parasol is more common than the parasol. It prefers shady places, and can be found in woods and shrubby areas and on garden compost heaps during summer and fall. The creamy buff cap, coated with curly brownish-red scales, ranges from 2 to 6 inches across. The thick, smooth stem is whitish with pink-brown tinges and has a thick, movable ring. The base is large and bulbous.

When collecting this mushroom, take care to avoid confusion with *Chlorophyllum molybdites* (see above). Some severe allergic reactions to the shaggy parasol have also been reported.

Collecting and Storing
Gather only firm young specimens; cut away the earthy bases before putting your find in the basket. This fleshy mushroom will stay fresh up to a week in the refrigerator. It is also suitable for drying.

Preparation and Cooking
Discard the tough stem, which is inedible. Wipe or wash the cap to remove any surface dirt.

The shaggy parasol can be prepared in the same way as the parasol mushroom: stuffed, made into fritters, fried and eaten for breakfast, or served in cream sauce. It is particularly good baked with Madeira. The large caps can be used to excellent effect as a base or "nest" for quail eggs with bacon, chicory and chives.

Parasol mushroom (Lepiota procera)

Shaggy parasol (Lepiota rhacodes)

1/3 life size

Oyster Mushroom

The Oyster Mushroom *(Pleurotus ostreatus)* grows on rotten stumps and fallen trunks of deciduous trees virtually all year, failing to appear only when the weather is very cold. This well-known edible species is now grown commercially and sold in markets (see below)—but it's more fun to collect your own!

Oyster mushrooms grow in clusters of overlapping tiers—like tiles—on decaying wood. As the common name suggests, the cap is shell-shaped and measures 2-1/2 to 5-1/2 inches across. Color varies greatly, ranging from cream to fawn to gray-brown or deep slate gray-blue. The gills are white, discoloring yellow and run down into a short, solid stem.

Collecting and Storing
This mushroom is well worth collecting. Besides being tasty, it has the added attraction of growing in large clumps, so you can usually pick enough for a decent meal. Only pry really young clusters from branches or stumps, since oyster mushrooms get very tough as they mature.

You can dry these mushrooms, but it is preferable to eat them fresh.

CULTIVATED OYSTER MUSHROOM
Commercial farming of oyster mushrooms has been underway for several years. It's a secretive business, and if you visit a grower, he won't let you into his growing sheds for fear that you might reveal some of his dark secrets. Kits are available, though, so it's possible to experiment with growing your own.

Oyster mushrooms are not grown on elaborately prepared trays of manure, but on roundish cushions of compacted straw, sawdust and compost encased by plastic mesh. Farmers have tried to simulate the mushrooms' natural growing conditions by placing the discs on top of each other; when the mycelium, a white thread-like mass, spreads over the surface, the mesh is removed to allow the mushrooms to grow out from the sides and form their true fan shape.

The cultivated mushrooms often have longer stems than their wild counterparts, and mature specimens tend to become funnel-shaped.

Farmers are nurturing the fawn and slate gray-blue varieties; there's also a pink variety and a new yellow type (labelled Yellow Pleurote) which loses its wonderful color when cooked.

Buying and Storing
Cultivated oyster mushrooms are widely available. Because this fragile species is easily broken, it is in many ways preferable to buy prepacked rather than loose mushrooms; those in baskets have a far better chance of remaining whole.

Always choose firm young specimens. They will stay fresh 3 to 4 days in the refrigerator and may also be dried or powdered.

Preparation and Cooking
Treat wild and cultivated specimens in exactly the same way. You can eat the whole mushroom—just wipe with a damp cloth to remove any dirt.

Oyster mushrooms can be eaten raw, but some people find them rather indigestible uncooked. They can be used in place of any of the cultivated mushrooms in most recipes; they're good fried, grilled or baked.

Deep-fried in batter, oyster mushrooms make crunchy beignets to dip into a thick sour cream and lemon balm sauce (see page 70). Like most fungi, they are delicious cooked with Madeira and cream, then served in puff pastry or tartlet shells or as a sauce for pork tenderloin. Or poach them whole in wine with Dover sole, saving a few to add to the sauce made from the juices.

Oyster mushroom (Pleurotus ostreatus)

Cultivated oyster mushroom

Yellow pleurote

Russulas

Some members of this genus are extremely colorful: you may find bright red, purple, chrome-yellow or green types in addition to the more familiar brown-hued ones. This ought to make identification easier—but there are so many russulas and each species is so variable, that it's hard to distinguish one from another. There's another distinctive characteristic that will help you recognize the genus, though: the brittle stem snaps like a piece of chalk.

Four popular edible russulas are illustrated here, but if possible, have an expert check your identification—it is better to be overcautious and healthy than too confident and ill. Some of the mushrooms in this group taste acrid and unpleasant, though none of them is seriously poisonous.

RUSSULA CYANOXANTHA

Russula cyanoxantha is a common species found, usually in deciduous woods, in summer and fall or, in California, in late fall and winter.

The flat cap can reach 6 inches across, tending to become depressed at the center as it matures. The color may be steely blue, gray, purple, greenish or a mixture of shades. The stem is white and firm. This fleshy russula has a mushroomy smell and mild, nutty taste.

RUSSULA CLAROFLAVA

A colorful mushroom, *Russula claroflava* is relatively easy to identify. It is found in woods in summer and fall.

The shiny bright yellow cap, 1-1/2 to 4 inches across, fades with age; it is rather sticky, and you should be able to peel the skin halfway back. The gills are cream-yellow; the stem is white, bruising gray when touched. The aroma is strong and fruity, but the flavor is mild. *Russula claroflava* adds color to most recipes and is ideal to mix with other species.

RUSSULA VIRESCENS

Connoisseurs consider *Russula virescens* among the best-tasting members of the genus. It is found in the woods of eastern North America during

1/2 life size

Russula cyanoxantha

Russula claroflava

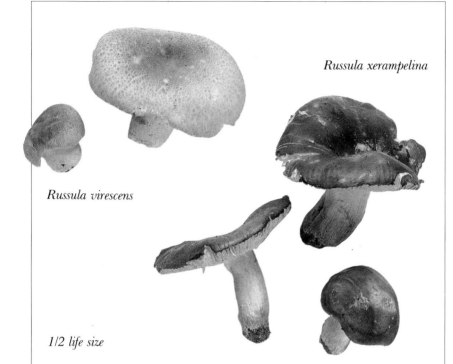

Russula xerampelina

Russula virescens

1/2 life size

summer and fall.

This russula has a dry, dull gray-green, blue-green or olive-green cap, 2 to 4 inches across; the skin tends to crack into small scales, showing white below. The short, fleshy white stem browns slightly with age. The mature cap has a mild nutty smell and flavor, but young mushrooms taste more like potatoes—which makes them a particularly suitable addition to potato recipes.

RUSSULA XERAMPELINA

Russula xerampelina tends to grow in mixed woods from midsummer until late fall. It has a dry, slightly depressed cap, 2 to 4-1/2 inches across, in varying hues of pink, purple or brown with a blackish center. The stem is white or rose-tinted.

This russula's distinctive crablike aroma, which persists even after cooking, makes it especially suited to seafood dishes.

Collecting and Storing Russulas

Squirrels have a penchant for the col-orful caps, while grubs like the fleshy stems, so collect young buttons before the woodland fauna have had their share. Russulas are rather brittle and should be placed in a basket to prevent them from breaking.

These mushrooms are best eaten fresh, but you can also preserve them in vinegar (see page 11).

Preparing and Cooking Russulas

If your russulas are fresh and young, use the whole mushroom, but discard the stalks of older specimens. If any leaves have stuck to the caps, wash them off under cold running water.

Russulas retain their firm texture when cooked. You can grill them, bake them or stuff them with toasted pecans, seasoned bread crumbs, shallots and cheese. A mixture of russulas gently fried in butter with onions, garlic and seasoning is delicious on toast. Russulas are also lovely served in cream sauce over pasta.

Russula virescens can be eaten raw; toss it in green salads or marinate it.

St. George's mushroom (Tricholoma gambosum)

St. George's Mushroom

St. George's Mushroom (*Tricholoma gambosum*) is better known in England than in North America. To an Englishman, the name implies that the species should first appear on St. George's Day, April 23—but unless the year is exceptionally warm, the mushrooms usually don't show up in Britain until a couple of weeks later.

St. George's mushroom bears a slight resemblance to the cultivated mushroom. Its thick, fleshy cap, 2 to 6 inches across, has an in-rolled margin; the color is white, discoloring to cream or apricot. The gills and stem are white. This mushroom's distinctive mealy smell and taste come out fully in soups.

Collecting and Storing
Only pick fresh young buttons, because the large old specimens are rather overpowering in flavor. These mushrooms will keep in the refrigerator a day or two. They're not really worth drying, since they lose their flavor.

Preparation and Cooking
St. George's mushrooms can be substituted for cultivated types in most recipes. Cut off the earthy bases and wipe the caps with a damp cloth to remove any dirt.

To prepare, simply slice and fry in butter with pepper, salt and a dash of lemon juice. Serve on toast for a light snack, or use as an omelet filling. To enrich this recipe, add garlic, sour cream and fresh herbs; use as a savory filling for puff pastry shells, to serve warm or cold with apéritifs.

To bring out the full flavor of St. George's mushrooms, use them to make a soup thickened with cream and egg yolks.

Field Blewit

The Field Blewit *(Clitocybe saeva)*, also called Blue Leg, is one of the best edible species—far superior to the Blewit (below)—but it is difficult to find. It is one of the few wild mushrooms to be eaten in Britain, especially in the Midlands, where it is sometimes sold at local markets. It is rare in North America.

You find field blewits in old grassy pastures, often growing in large rings; they're easy to miss, since their dull gray-brown, wavy-edged caps look like shrivelled leaves wedged between the grass stalks. The caps measure 2 to 4 inches across. The gills are flesh-colored; the creamy stem has a distinct bluish-mauve tinge, hence the common name "blewit." The stem base is sometimes swollen.

Field blewits can tolerate a certain amount of cold weather and appear during fall and early winter. They have a strong perfumy smell and taste.

Collecting and Storing
Pick these mushrooms on a dry day; the porous caps become slimy in wet weather. They will keep a day or two if refrigerated.

Preparation and Cooking
Use the whole mushroom, first removing dirt with a damp cloth. This fragrant mushroom is superb combined with chicken in a white sauce. You can also fill the large caps with a savory stuffing made with the diced stalks, then bake them.

Blewit

A bluish-lilac mushroom which turns brown with age, the Blewit *(Clitocybe nuda)* is common in gardens and mixed woodlands during fall and early winter. In England, it is known as the Wood Blewit. Unlike its cousin the field blewit (above), it has purple gills which fade to a light brown. The cap measures 2-1/2 to 4-1/2 inches across. Fresh blewits have a sweet, perfumy smell, but after cooking they taste much like new potatoes.

Collecting and Storing
This mushroom can be kept in the refrigerator a day or two.

Preparation and Cooking
Because blewits are slightly toxic and may cause an allergic reaction, they should not be eaten raw. Parboil them before using them in recipes. Cook as for field blewits; add to potatoes *au gratin*.

Blewit (Clitocybe nuda)

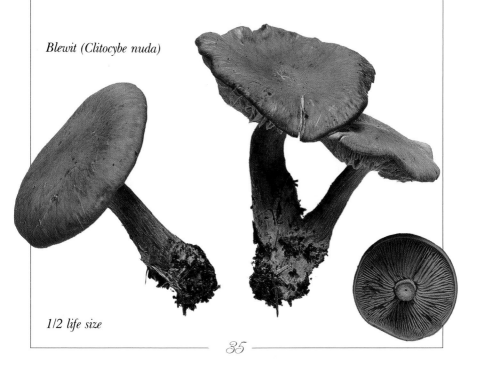

1/2 life size

1/2 life size   *Cepe (Boletus edulis)*

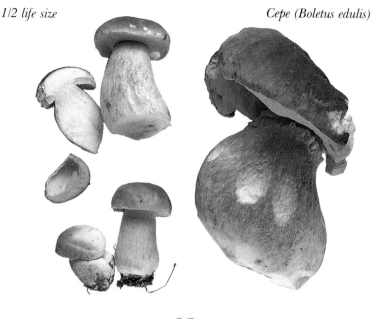

Cepe

The Cepe *(Boletus edulis)* belongs to the boletes, a group of mushrooms which have tubes—a spongy layer covered with tiny pores—instead of gills.

King of the edible mushrooms, the cepe is widely distributed in North America, particularly in the West; it grows in the woods during summer and fall or, in California, in fall and winter. It is avidly searched for in Europe—especially in Italy, where it is collected in vast quantities and dried for the winter.

The French call this mushroom cepe ("vine stock") because its fat stem looks like a trunk; in Britain, it's known as the Penny Bun because the sticky brown cap resembles a bread roll. I have yet to fathom why the Italians refer to it as *porcini*, which translates as "piglets!" Whatever the name (and the spelling—in many cookbooks, you'll see the spelling *cèpe*), this mushroom has a delicious nutty flavor.

The cepe has a brownish cap, 3 to 8 inches across, which becomes viscid in damp weather. It has cream to olive-yellow pores and a thick, bulging fawn stem partially covered with a fine network pattern.

Collecting and Storing
Pick fresh specimens with firm caps; these will keep in the refrigerator about a week. Cepes are excellent for drying, as they retain their flavor and a small amount goes a long way.

If you can't find any cepes growing wild, you can sometimes buy them fresh from specialty markets. The dried form (usually labelled porcini) is available all year in delicatessens and gourmet food shops. Old porcini blacken, so buy only cream and brownish ones which have been recently imported.

Preparation and Cooking
These mushrooms are much sought after by grubs as well as people, so it's wise to cut them in half or remove the stalk to check for infestation. Clean the caps with a damp cloth; scoop away the pores if they have turned yellowish-green.

Cepes can be cooked in numerous ways. Fry them in olive oil and add crushed garlic and parsley just before serving; or stuff the large caps with ham or bacon, shallots, herbs, bread crumbs and grated Parmesan cheese, then brush with a little oil and bake. Cepes can be made into pasta sauces, and the Italians fill ravioli with them.

The young caps are delicious eaten raw in green or seafood salads.

If you have any uncooked cepes leftover, preserve them in an herbed marinade for later use—they're excellent as an hors d'oeuvre.

Bay Boletus

Another delicious bolete, the Bay Boletus *(Boletus badius)* is common in eastern North America in summer and fall. It grows in coniferous woods, but can occasionally be spotted under deciduous trees.

This mushroom is smaller than the cepe, and its slender stem— white at the top, with brown markings below—lacks the cepe's fine network pattern. The felty, red-brown to chocolate-brown cap, 1-1/2 to 5-1/2 inches across, becomes rather clammy when wet. An identifying feature of bay boletus is that the lemon-yellow pores bruise blue-gray when touched. Both taste and smell are mild and mushroomy.

Collecting and Storing

Gather young specimens with firm caps. They will stay fresh a week if refrigerated and, like cepes, are ideal for drying.

Preparation and Cooking

This species is not a favorite of woodland grubs, so there's no need to remove the stalk. Just clean the mushrooms with a damp cloth.

Cook as for cepes (see opposite) or, if you have only managed to find a few specimens, use to flavor sautéed potatoes. Or cook in cream and use to fill an omelet.

Red-cracked Boletus

The Red-cracked Boletus *(Boletus chrysenteron)* is found in woods throughout North America, but it is not one of the best wild mushrooms to eat.

This bolete's common name refers to its distinguishing feature: the velvety tan to olive-brown surface of the cap cracks to reveal yellow or coral flesh beneath. The cap measures 1-1/2 to 4-1/2 inches across. The pale yellow pores turn greenish with age; the stem is yellow at the top, with a rosy-red base. The cut flesh slowly turns bluish; it has a mild smell and taste.

Collecting and Storing

This mushroom is not really suitable for cooking, since it turns mushy. It is worth collecting for drying, though, as it retains its flavor well. Pick firm young specimens.

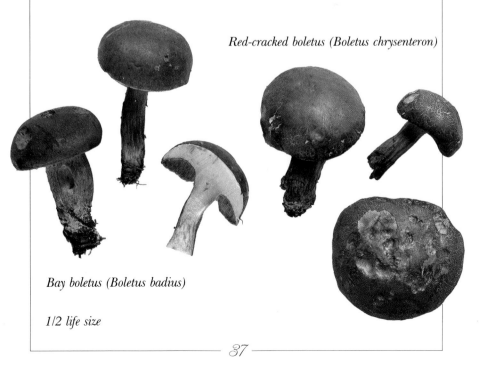

Red-cracked boletus (Boletus chrysenteron)

Bay boletus (Boletus badius)

1/2 life size

Birch Boletus

All members of the genus *Leccinum* have dry caps and woolly, scaly stems; they are often called "rough stalks." All are edible.

The Birch Boletus *(Leccinum scrabrum)* is the most prolific of these tall-stemmed boletes; unfortunately, however, it's not particularly good to eat. It can be found growing in grass under birch or aspen trees throughout the summer and fall.

The velvety cap, 2 to 6 inches across, is hazel to snuff-brown and turns slimy when wet; the cream-colored pores bruise gray. The firm white stem is covered with rough, dark brown scales. The flesh is mild in flavor and aroma, and does not change color when cut.

Collecting and Storing
Gather only firm young specimens; the mature caps are spongy. The best way to store this mushroom is to dry and powder it.

Preparation and Cooking
The birch boletus is often maggoty, so use only the caps. If the pores feel squashy, discard them; then wipe the caps with a damp cloth. These mushrooms are suitable only for making soup.

Leccinum quercinum

This rare mushroom grows only under oaks. It has not been described in North America, but if you happen to live in Southern England, you may be lucky enough to find it from midsummer until mid-fall.

Leccinum quercinum has been nicknamed the Brick Cap, thanks to the brick-red color of its 2-1/2- to 6-inch-wide cap. The pores are buff, becoming darker with age; the thick stem is coated by a fine network of orange scales. The flesh turns pink-gray when cut. Both scent and flavor are pleasant.

Collecting and Storing
If you find one *Leccinum quercinum*, it's worth searching around beneath neighboring oaks for more: though these mushrooms appear only singly or in pairs, they can be scattered over a large area. *Leccinum quercinum* is a robust species, fairly bug-free, so it is safe to collect mature specimens. Like its cousins, it is excellent dried.

Preparation and Cooking
Clean with a damp cloth and cut away the stem bases. Cook as for cepes (see page 36).

Orange Birch Boletus

Recognizable by its striking orange or yellowish-red cap, the Orange Birch Boletus *(Leccinum testaceoscabrum)* commonly grows under birch. It can be found throughout the summer and fall.

This is a tall mushroom with a cap measuring up to 8 inches across. The pores are gray at first, turning to white and then dirty yellow, bruising blue-gray. The stem, often club-shaped, is covered with woolly brown scales. The flesh blackens when cut and cooked, but don't be put off by this—orange birch boletus is a good edible species.

A close relative, *Leccinum aurantiacum* has a downy, apricot-colored to brown cap, 3 to 6-1/2 inches across, and a tall stem covered with rusty-brown scales. The cut flesh turns pinkish-gray and gradually blackens. This mushroom and other near relations, notably *Leccinum insigne,* associate with aspen and are good edibles. They may be found during summer and fall.

Collecting and Storing
Pick mushrooms with firm caps. The orange birch boletus retains its flavor well when dried and powdered; in that form, it's ideal for use during the winter, as a seasoning for soups, terrines or mushroom dishes.

Birch boletus (Leccinum scabrum)

Orange birch boletus (Leccinum testaceoscabrum)

Leccinum quercinum

1/2 life size

Preparation and Cooking

The fibrous stems are too tough to eat, so use only the caps. If the pores have turned soggy and dirty yellow in color, discard them; they are easy to detach and peel away from the cap. Wipe the caps with a damp cloth.

The large caps are ideal for stuffing and baking. Make the small caps into a sauce to serve with pasta; or, if they are firm and saucer-shaped, pour a few drops of olive oil into the center of each, add some fresh herbs, crushed garlic and a dash of lemon juice, wrap in foil and cook over a barbecue. If you have only a few orange birch boletus, mix them with other mushrooms, preferably cepes.

Suillus

Most members of this genus of boletes are found in association with conifers. The majority have shiny, sticky caps, and all have pores and spongy flesh instead of gills.

Suillus are flavorful but not good to eat fresh: they soak up cooking fat or oil like sponges and become soggy and slimy. Nonetheless, they are well worth gathering for drying; you can use them as a flavoring for soups, sauces and stews.

SLIPPERY JACK

Slippery Jack (*Suillus luteus*) is a large member of the genus, found in abundance in conifer forests during summer and fall. Its chestnut-brown cap, 2 to 4-1/2 inches across, is coated with a shiny lilac-tinted glutin; the pores are lemon-yellow, the stem dirty yellow with a large, floppy brownish-purple ring.

Slippery jack is a good edible fungus, but it's prone to insect attack, especially during warm weather; you'll find better specimens if you pick it later in the season. Its taste and smell are not distinctive.

SUILLUS BOVINUS

Suillus bovinus is common in coniferous woods in England, especially in damp areas, during summer and fall. It is not found in North America. Both cap and stem are pinkish-tawny; the sticky cap, 1-1/4 to 4 inches across, has a distinct white margin. The pores are large and greenish-yellow. *Suillus bovinus* has a fruity smell and a pleasant flavor.

SUILLUS VARIEGATUS

In England, this species can be found in mixed coniferous forests during summer and fall. It resembles *Suillus tomentosus*, found in the fall in North America. The 2-1/2- to 5-inch-wide cap and firm stem are mustard-colored; the pores are olive-green tinged with blue. The cap has a felty surface speckled with rust to dark brown scales; it becomes sticky in wet weather. Though it smells strong, this mushroom has a mild taste.

LARCH BOLETUS

The Larch Boletus (*Suillus grevillei*), also known as *Boletus elegans*, grows exclusively under larch trees and can be harvested in large quantities during summer and fall. The tacky golden cap, 1-1/4 to 4 inches across, darkens to a rusty orange and turns shiny as it matures and dries. The lemon-yellow pores bruise brownish when touched; the orange-tinged yellow stem has a whitish ring. Both odor and flavor are mild.

SUILLUS GRANULATUS

Suillus granulatus is common on sandy soil in coniferous woods throughout summer and fall. Its glutinous reddish-brown cap, 1-1/4 to 3-1/2 inches across, is easy to peel and turns shiny when dry. The lemon-colored pores exude tiny, milky droplets. The upper section of the whitish to pale yellow stem is covered in granules, hence the specific name.

Collect only fresh young specimens; the older ones tend to be attacked by maggots. Like *Suillus bovinus*, this species has a fruity smell and tastes pleasant.

Collecting and Storing Suillus

Pick fresh young specimens; cut off the earthy bases before putting the mushrooms in your basket.

Members of the genus *Suillus* are suitable only for drying and using as a flavoring. To preserve them, you must first remove the gelatinous skin—a somewhat tricky process. The best method is to wash the caps under cold running water, then place them on paper towels and let them dry 30 minutes. As you pull the paper away, the skin will come with it.

Cut the caps and stalks into thin slivers, then dry them on a baking sheet in a low oven (see page 11). Store airtight.

Uses

To flavor pureed soups, soak the dried slices in water 20 minutes to soften, then add them—along with their soaking liquid—to the stock you're using. They're a delicious addition to all vegetable soups.

You can also grind the dried mushrooms to a powder.

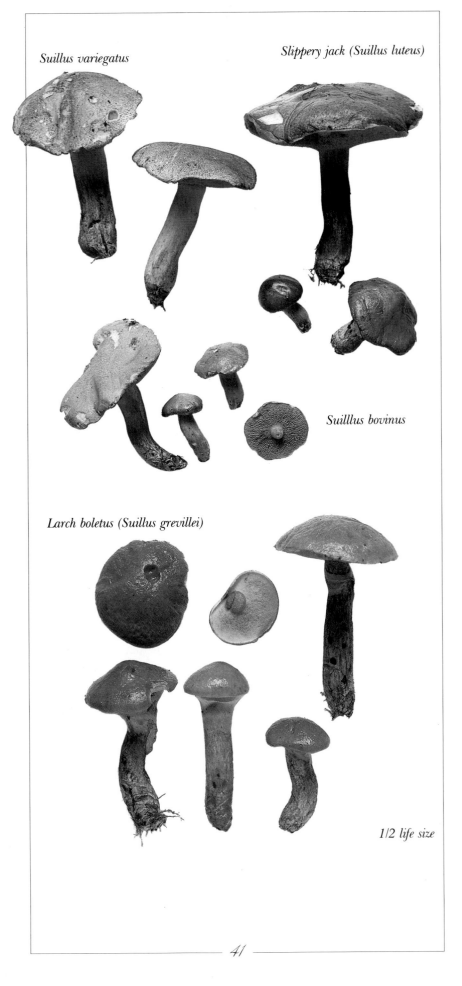

Suillus variegatus

Slippery jack (Suillus luteus)

Suilllus bovinus

Larch boletus (Suillus grevillei)

1/2 life size

Caesar's Mushroom

Though there are a number of good edible mushrooms in the genus *Amanita*, the group also contains several deadly species. For this reason, all beginners should stay away from the amanitas; in fact, even some experienced mushroom-hunters eschew them entirely. An essential skill for any hunter is knowing how to avoid picking them inadvertently!

Despite its lethal relatives, Caesar's Mushroom *(Amanita caesarea)* is very popular in France—and was the most highly prized edible species of ancient Rome. In North America, it is found in the East and as far west as Arizona and New Mexico. Its striking bright orange cap, measuring up to 7 inches across, fades to yellow with age. The gills and club-shaped stem are yellow; the stem has a drooping ring which is easily detached. The base of the stalk is encased in a white volva (a sacklike structure). Caesar's mushroom has a faint smell and tastes pleasant.

Other edible members of the genus include the Blusher *(Amanita rubescens)*, the Tawny Griselle *(Amanita fulva)* and the Coccora *(Amanita calyptrata)*, all quite common in the woods. Unless you are *totally* familiar with them, however, you are strongly urged to leave them strictly alone: they have a number of deadly relatives, among them the Death Cap *(Amanita phalloides)* and the Destroying Angels *(Amanita verna, Amanita virosa, Amanita ocreata* and *Amanita bisporigera)*. Another relative, *Amanita pantherina*, contains different toxins which, in quantity, have also been reported to be fatal. Other species may make you seriously ill.

The genus *Amanita* has certain identifying characteristics. When very young, the mushrooms look egglike and are encased in a veil through which the buttons burst; the cap is normally covered with flecks of veil remnant similar to tiny warts, but heavy rain can wash these away. The stalk is encased in a volva, which may be loose and baglike or show only as a rim around the base. The gills are typically white, sometimes yellow, and discolor with age. A white spore print is a more reliable indicator than gill color (a field guide can tell you how to take a spore print).

Collecting and Storing

If you are travelling around the Mediterranean during summer or fall, you may spot Caesar's mushroom growing in an oak wood. Don't confuse it with the poisonous red Fly Agaric *(Amanita muscaria)*.

If you find a Caesar's mushroom, use a knife to dig out the whole fruiting body; be sure to collect the stem from under the ground so the volva can be seen. Take it to an expert or local pharmacy to have your identification confirmed; most French pharmacists have a display of edible fungi in their windows during the season, and they are only too pleased to help.

Use the mushrooms the day they are picked.

I cannot emphasize enough that *extreme care must be taken when gathering any edible amanita.* Do not eat any amanitas until you have had their identity confirmed—if you get it wrong, you may die.

Preparation and Cooking

Wipe the caps and clean the stalks thoroughly. Caesar's mushroom is delicious gently fried in olive oil with shallots. Just before serving, add a little crushed garlic and some finely chopped parsley.

Note: None of these mushrooms is illustrated. For identification, it is essential to consult a field guide.

Hedgehog fungus (Hydnum repandum)

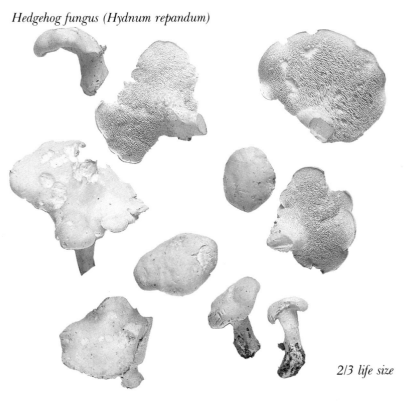

2/3 *life size*

Hedgehog Fungus

The Hedgehog Fungus (*Hydnum repandum*, also known as *Dentinum repandum*), differs from all other edible fungi because it has spines instead of gills or pores, hence the common name. It is one of the easiest species to identify, and there are no poisonous lookalikes to cause confusion. It is widely available in Europe and now appears in some stores in the United States.

The hedgehog fungus has a velvety, creamy-apricot cap, 1-1/4 to 6-1/2 inches across, which becomes slightly funnel-shaped and tends to fold in on itself. In this form, it looks something like a cloven hoof, thus the French name *pied de mouton* ("sheep's foot").

Hedgehog fungus is found among hardwoods and conifers from summer until fall. It has a pleasant smell; the flavor is rather bitter with a peppery aftertaste when the fungus is raw, but milder after cooking.

Collecting and Storing
Hedgehog fungus grows among leaf litter, and it is easy to pull the whole plant from the ground gently. Once you have done this, cut away the earthy base of the stem and remove all the old leaves nestling in the cap before placing your find in the basket; the less dirt you take home, the easier the fungus will be to clean.

These mushrooms will keep in the refrigerator 10 days. They are enjoyable to eat fresh, but not worth preserving for later use.

Preparation and Cooking
Wipe the caps with a damp cloth. If you were careful when you picked the fungus, there shouldn't be any soil trapped in the spines; if there is, try to dislodge it with a small brush.

Hedgehog fungi are rather dry and require slow cooking over low heat. They are suitable for use in most mushroom recipes. Fried in butter, flavored with fresh herbs and served on toast, they make a tasty snack. They are particularly good made into a sauce and served with chicken or game, or added to a stew.

Morel

Mushroom connoisseurs consider Morels among the finest edible fungi, and fresh specimens are much sought after when they come into season in spring. They are found throughout North America and, though rare in Britain, are abundant in most European countries— especially in France, where vast quantities are collected for drying or canning. Recently, techniques have been developed to make commercial cultivation practical. When fresh, morels have a sweet, earthy smell and a slightly nutty taste.

MORCHELLA

The Common Morel (*Morchella vulgaris*) ranges from 2 to 5-1/2 inches tall and resembles a natural sponge on a stalk. The hollow, dome-shaped cap is gray-brown, turning sandy with age; the finely pitted surface resembles a honeycomb.

Common morel is not reported in North America. In Europe, it can be found growing in rich soil in gardens and open woodland during late spring.

MORCHELLA ESCULENTA

Morchella esculenta is 2-1/2 to 8 inches high—slightly taller than its cousin, the common morel. The cap profile varies from bell-shaped to globular; the honeycomb surface is camel-colored, browning with age. Both the cap and the furrowed, whitish stem are hollow.

These morels apparently favor disturbed habitats; in particular, they often appear in the spring following a forest fire. Thus, in Yellowstone National Park, the huge fires of 1988 were followed by abundant fruitings that continued throughout the summer of 1989. Legend has it that peasant women in 18th-century Germany deliberately started forest fires to persuade these mushrooms to grow. And old records reveal that morels grew in groups on bomb sites in England during World War II.

Besides *Morchella esculenta*, Europeans also enjoy eating the closely related *Morchella conica*, a small species with a dark gray, cone-shaped cap, and *Morchella rotunda*, a spongy, beige-capped mushroom which is the largest member of the family. Several other species may be found in North America, including the White Morel (*Morchella deliciosa*) and the Black Morel (*Morchella elata*).

Collecting and Storing

If you are fortunate enough to find a colony of morels, cut off their earthy bases before you put them in the basket to keep grit from getting trapped in the pitted cap surface.

All morels are excellent for drying: thread the halved and cleaned mushrooms on fine string and hang them above a radiator or in a warm kitchen for a couple of days. When they are dry and crisp, put them in an airtight jar. To reconstitute, soak them in warm water 30 minutes.

If you would like to taste this mushroom but cannot find any fresh species, you can buy packaged dried morels from gourmet shops and some delicatessens.

Preparation and Cooking

Cut each mushroom in half and wash carefully under cold running water to remove any dirt or insects which may be hiding in the crevices.

Morels should never be eaten raw, since they are liable to cause nasty stomach upsets; in fact, they have been known to cause allergic reactions even when cooked. It's a good idea to blanch them a minute or two before cooking.

Morels are exquisite prepared with cream, then eaten on their own as a vegetable dish or served as a sauce for veal; if you add a few spoonfuls of Madeira, they're lovely with chicken. Or combine them with other species to make a soup.

Like all mushrooms, morels go well with eggs. They are delicious with scrambled eggs and in omelets or quiches.

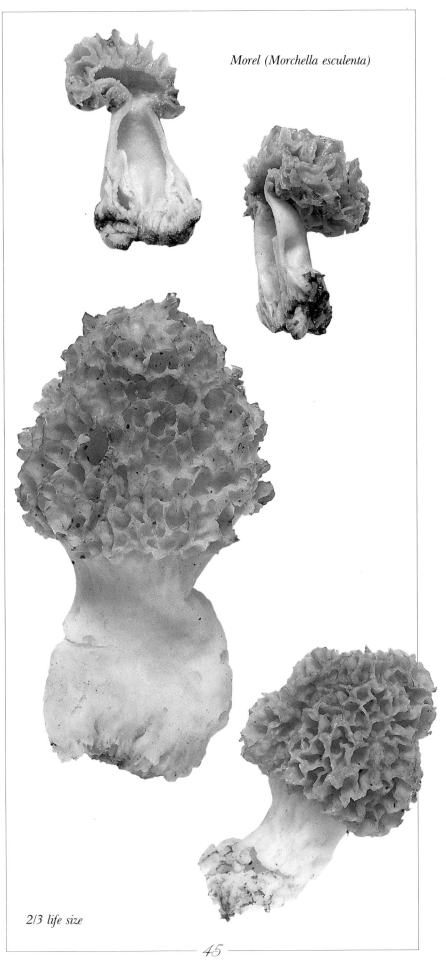

Morel (Morchella esculenta)

2/3 life size

3/4 life size

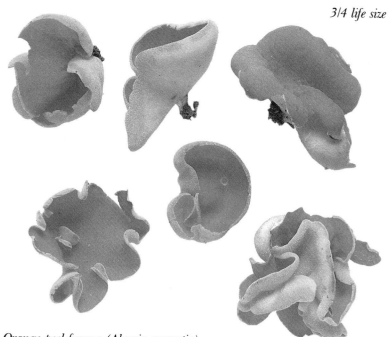

Orange peel fungus (Aleuria aurantia)

Orange Peel Fungus

Brightly colored and very common, the Orange Peel Fungus *(Aleuria aurantia)* is a cup fungus usually found growing in bare gravel, on paths or among grass in lawns and on roadsides from early fall until the beginning of winter.

The cup, often with a split down one side, measures 1/2 to 4 inches across; it becomes flattened and wavy with age. The inside of the cup is bright orange, but the outer surface is coated with a fine whitish down.

Collecting and Storing
Use a knife to pry the mushrooms out of the ground; cut away the earthy bases before putting them into your container to take home. I have never found orange peel fungus in large enough quantities to preserve, but it can be dried, though its beautiful color fades to a rather dull brown.

Preparation and Cooking
Brush off any surface grit and wipe with a damp cloth. This fungus's flavor is extremely mild and delicate— hardly discernible, really— but its brilliant color makes it an attractive embellishment for other mushroom dishes, soups, salads or green vegetables (particularly broccoli). If you have found only a small number, cut them into thin slivers; otherwise, use them whole.

Scarlet Elf Cup

A brilliant red cup fungus, Scarlet Elf Cup *(Sarcoscypha coccinea)* is quite rare in comparison to its cousin the orange peel fungus. It is found on dead wood during winter and early spring. The cup measures 1/2 to 2 inches across; the inside is scarlet, while the outside looks as if it had been dusted with a fine white powder. Scarlet elf cup is very similar to orange peel fungus and makes a striking garnish for many savory dishes.

Cloud Ear

Cloud Ear *(Auricularia auricula)* can be found throughout the year, but is most prolific during fall. It is an ear-shaped bracket fungus, 1-1/4 to 3-1/4 inches across, which grows in clusters and folds on dead branches of broad-leafed trees. The upper surface is a soft, velvety reddish-brown; the inside is smooth and dull brown, with a jellylike feel. With age, cloud ear turns dark brown and becomes rock hard.

This rather ugly fungus is a good edible species that is easy to identify. It is highly regarded in China; along with related species such as Wood Ear (see page 56), it is cultivated and dried, then sold in Asian grocery stores throughout the world. In the past, cloud ears were much esteemed by herbalists, who used them in poultices to soothe sore eyes.

Collecting and Storing

Cloud ears should be gathered when young and soft. Use a knife to cut them from the trees. Separate the clusters and wash well under cold running water.

Cloud ears will stay fresh in the refrigerator about 2 days. You can dry them, but they lack the flavor of their Chinese relatives.

Preparation and Cooking

Cloud ears need long, slow cooking—even if they feel soft, they have a tendency to become leathery, especially if they're on the elderly side. Cut the fungus into into thin strips and simmer in milk 40 minutes; then reduce the liquid and season with salt and plenty of pepper. Or simmer gently in butter with fresh herbs and serve on toast.

Cloud ears can be combined with other Asian species and used in most Chinese dishes, especially those including pork.

Cloud ear (Auricularia auricula)

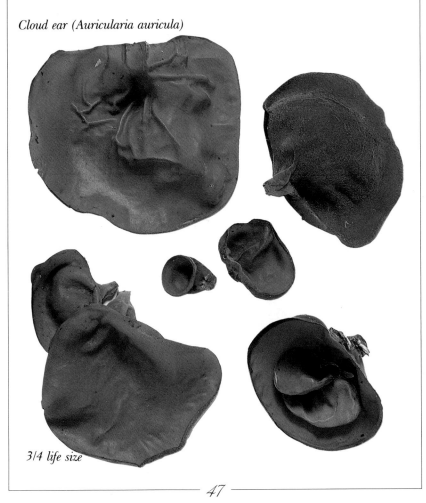

3/4 life size

Beefsteak fungus (Fistulina hepatica)

1/2 life size

Beefsteak Fungus

The Beefsteak Fungus *(Fistulina hepatica)* takes its name from its looks: the cut flesh looks like prime beef, and a reddish juice seeps from the slices. It is tonguelike in shape, hence its other name, Ox Tongue.

This species is a bracket fungus, sometimes as large as 12 inches across, which grows on the trunks of trees—occasionally on chestnuts, but usually on large oaks—during late summer and fall. It is not common in North America except in the Southeast. The upper side is reddish-orange, darkening to purple-brown with age. The spores are pinkish-brown and bruise a dull red when touched. The smell is pleasant, but the flavor of the uncooked fungus is rather sour.

Beefsteak fungus causes brown rot in wood; infected oak acquires a darker, richer color and is highly prized among furniture makers, who refer to it as "brown oak."

Collecting and Storing
Use a knife to pry the fungus away from the bark. Collect only small, tender specimens, since this species becomes tough and very bitter as it ages.

Though beefsteak fungus will keep 2 days in the refrigerator, ideally it should be eaten as soon as possible; it has a tendency to dry out and become leathery.

Preparation and Cooking
Since this fungus prefers to grow midway up an oak trunk, it is usually very clean and only requires wiping with a moist cloth.

The best way to cook beefsteak fungus is to cut it into thin slivers and sauté slowly in butter with shallots, garlic and fresh herbs, especially thyme or marjoram, about 20 minutes. The fungus releases a vast quantity of juice, so increase the heat and reduce some of the liquid before serving.

Eat as an appetizer, with plenty of French bread to soak up the delicious sauce.

Dryad's Saddle

A large bracket fungus, Dryad's Saddle *(Polyporus squamosus)* is considered worth eating, but old, fibrous specimens are tough and unpalatable. Usually fan-shaped and sometimes up to 24 inches across, dryad's saddle can be found growing on deciduous trees, especially elm and beech, from late spring through fall.

These mushrooms tend to overlap one another as they grow. The fawn-brown upper surface is speckled with darker scales; the spores are cream-colored. The odor is strong and mealy.

Note: This mushroom is not illustrated.

Chicken of the Woods

Found during spring and summer, Chicken of the Woods or Sulfur Shelf *(Laetiporus sulphureus)* is a fleshy bracket fungus which grows in large tiers on a variety of trees. Its color ranges from bright yellow to orange, fading to buff with age; its wrinkled surface feels like chamois leather. It measures up to 16 inches across and is irregularly fan-shaped or semicircular. If you squeeze a young fungus, its thick flesh exudes a lemon-colored juice, but older, drier specimens will simply crumble.

This mushroom is considered a delicacy by many. But because a number of toxic reactions have been reported, it should be treated with caution, particularly when found growing on eucalyptus. It is rarely eaten in England but is popular in Germany.

Chicken of the woods has a strong smell, a pleasant taste and a fibrous, meaty texture. It is often used in place of chicken, hence its common name.

Collecting and Storing

Collect only young specimens by cutting them away from the tree bark. Chicken of the woods will keep in the refrigerator up to 2 days and, unlike most fungi, is also suitable for freezing. Before freezing the fungus, cut it into small cubes; then either blanch or fry in butter.

Preparation and Cooking

If you have gathered young specimens, you can use virtually the whole bracket, removing only the woody core. Use only the tender, pale yellow, frilly edges of the larger tiers. Clean by wiping with a damp cloth, then slice.

Chicken of the woods is wonderful added to chicken casseroles or made into a sauce; it complements all chicken recipes and can even be used as a substitute.

Thin slivers sautéed in butter make an attractive garnish for soups and salads.

The simplest and possibly the most delicious way to cook this mushroom is to fry it gently in butter with seasonings and finely chopped shallots; just before the mushrooms are done, add a generous squeeze of lemon juice. Serve alone or on toast for a light lunch. For an interesting starter, add a little whipping cream to the sautéed mushrooms, cook a few more minutes and garnish with parsley.

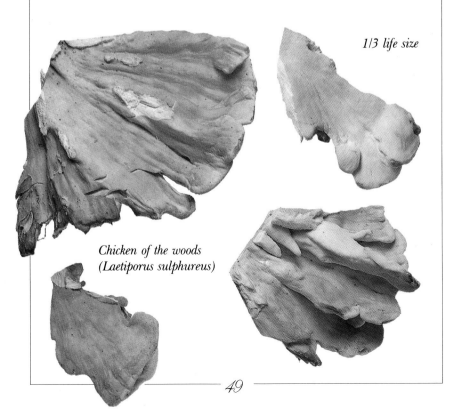

1/3 life size

*Chicken of the woods
(Laetiporus sulphureus)*

Cauliflower Fungus

The Cauliflower Fungus (*Sparassis crispa*) resembles the heart of a cauliflower; some claim it looks rather more like brains, which explains why it has also been christened Brain Fungus! It also bears a strong resemblance—in both shape and color—to a natural sponge. You can occasionally find it growing at the base of conifers, particularly pines, during fall.

This easily identifiable fungus may reach 20 inches across. Ocher-colored when young, it discolors with age. Young specimens have a slightly nutty taste and a sweetish aroma, but older ones become tough and bitter.

Few mushroom experts utilize this fungus because it takes so long to prepare, but you may find the unique, delicate flavor well worth the trouble. Cauliflower fungus sometimes appears in stores. If you ever spot what you think to be this species, examine it carefully—unless it's young and fresh, it's not worth eating.

Collecting and Storing

Pick only fresh young specimens. If you are successful in finding a cauliflower fungus imbedded in the roots of a pine, cut it from the thick stem with a sharp knife. It is quite fragile, so put it in a basket (rather than a bag) to keep it from getting crushed.

This fungus is usually so huge that even if you give half of it away, you'll have enough for several meals. If refrigerated, it will keep up to 10 days. It can be dried for use as a soup or stew flavoring. It also freezes well; you can blanch the flowerets, but it's preferable to sauté them gently in butter prior to freezing.

Preparation and Cooking

Cauliflower fungus has just one drawback: cleaning it is a laborious, time-consuming task. But don't let that deter you; the delicious flavor will be ample reward for all your work. Start by removing all the pine needles, pieces of bracken and dead leaves trapped in the frilly surface. Carefully break the fungus into flowerets and cut away any brown edges or spongy parts. Rinse under cold running water to remove any grit or small insects which may be hiding in the folds; it's a good idea to place a colander under the faucet to catch any pieces that break off.

A simple way to cook cauliflower fungus is to dust the flowerets with flour and gently fry in butter a few minutes; add seasoning, chives, parsley and thyme, and cook over low heat to draw out the natural liquid. Thicken the juices with egg yolk, then serve with a slice of lemon and a sprinkling of chopped herbs. If you prefer a piquant flavor, cook the fungus with spices instead of herbs.

A more traditional method of preparation calls for baking the fungus in a casserole with butter, parsley, finely chopped onion, a little garlic and some chicken stock.

Cauliflower fungus is good in most rice recipes, and it's an interesting addition to savory fillings for tomatoes, zucchini or vine leaves. It can be used on its own or mixed with cultivated mushrooms.

The nutty flavor of this fungus makes it pleasant to eat raw, and the pale yellow flowerets look attractive tossed in a green salad or added to a shellfish dish. If you are lucky enough to find cauliflower fungus, serve small pieces mixed with fresh crab in a dill vinaigrette.

3/4 life size

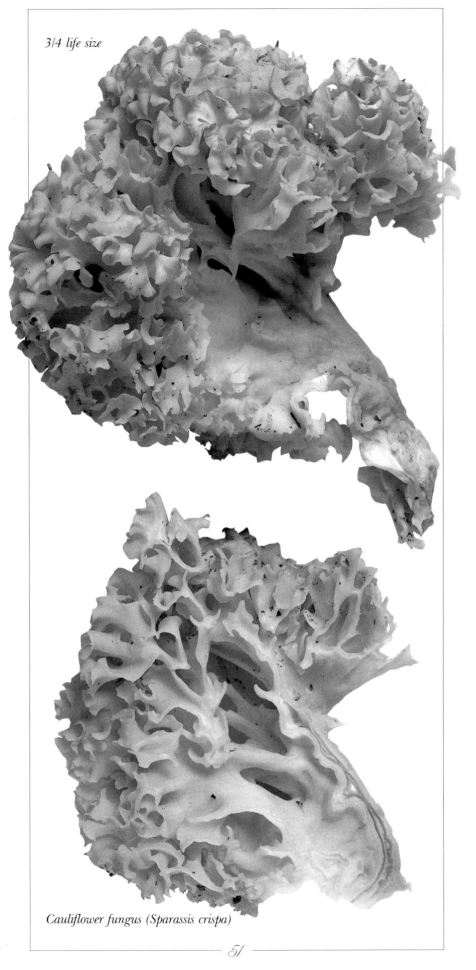

Cauliflower fungus (Sparassis crispa)

Giant Puffball

There is little point in searching specifically for Giant Puffballs *(Calvatia gigantea)*—if they're around, it's almost impossible to miss them. You stumble across them nestling among a clump of nettles or tucked underneath a hedge, or you spot them from miles off glistening on a facing hillside or in a stubble field, looking just like abandoned soccer balls. In fact, you must try to resist the temptation to kick them!

Because of their immense size (the girth can expand to 28 inches), giant puffballs are sometimes thought of as common, but you can go several years without seeing even one. Their season runs from late spring through fall. They are white at first, gradually fading to dirty yellow, then to brown when the leathery skin splits and the mushroom starts to disintegrate.

Collecting and Storing

If you are lucky enough to spot giant puffballs, collect only pure white ones. Try to keep each intact until you reach home, because it's fun to weigh and measure them—as you would prize pumpkins—and keep a record to compare with future finds.

A young specimen will keep in the refrigerator 2 days. You can freeze puffballs, but they tend to become rather soggy when thawed. If you have more than you can use, it's preferable to share your prize with friends and hope you're fortunate enough to find another one next season.

Preparation and Cooking

Cut away the dirty base and wipe the surface with a damp cloth. There is no need to peel the kid-leather skin from a puffball—just cut the unpeeled mushroom into about 1/2-inch-thick slices. The flesh has a spongy, marshmallowy texture.

A popular way to cook the slices is to dip them in an egg batter, then coat with toasted bread crumbs, fry in bacon fat until golden brown and serve hot for breakfast. Or add a handful of chopped herbs to the batter, cook in olive oil with some garlic, and serve as a light lunch.

Giant puffball (Calvatia gigantea)

1/3 life size

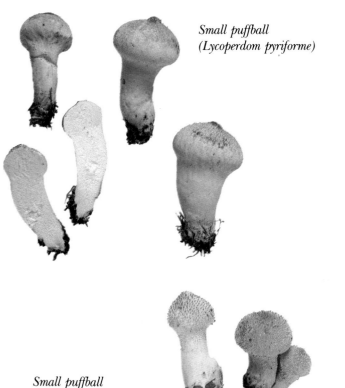

*Small puffball
(Lycoperdom pyriforme)*

*Small puffball
(Lycoperdom perlatum)*

1/3 life size

Small Puffballs

The edible Small Puffballs are thin-skinned and should not be confused with their thick-skinned and toxic cousins, the Earthballs. A much more serious mistake would be to collect the button stage of an *Amanita* (see page 42), which can look very much like a small puffball until its veil ruptures. To be safe, cut the purported puffball through its center; if you see the slightest indication of a gilled mushroom, discard your find. A genuine puffball is quite homogeneous inside its membrane.

LYCOPERDON PERLATUM
This common puffball, up to 3-1/2 inches high, resembles a household light bulb in shape. It is white at first, turning dirty gray and finally yellowish-brown as it matures. The flesh becomes powdery and disintegrates with age. This species can be found in all types of woodlands during late summer and fall.

LYCOPERDON PYRIFORME
Common from late summer through fall, this small, pear-shaped puffball has rough skin that feels like coarse sandpaper. It reaches about 1-1/2 inches high and is usually found in clusters on fallen trunks, old stumps, or rotting wood; though it sometimes may appear to be growing right out of the earth, it will in fact be attached to a buried log. Young specimens are whitish or light beige; older ones darken to cinnamon-brown.

VASCELLUM PRATENSE
This very common small puffball, cream in color, is found growing on lawns, golf courses and pastures throughout the fall.

Collecting and Preparing
Always collect young specimens. Cut away the earthy base to check that the flesh is pure white; if it has started to turn yellow, discard the mushroom. Most species have tough, scaly skin and require peeling before use.

Small puffballs can be cooked like giant puffballs (see opposite) or made into fritters.

White Helvella

The White Helvella *(Helvella crispa)*, a relative of the morel (see page 44), can be found throughout the fall and occasionally also appears in spring. It grows in moss and along the sides of paths in damp coniferous and deciduous woods. In a good year, you may come across as many as twenty growing in a line, but usually you'll find only two or three.

You can recognize this 3/4- to 2-inch-tall mushroom by its saddle-shaped cap—light cream on top, pale buff or tan beneath—which may be twisted and puckered like a piece of discarded paper. The stem is hollow and deeply furrowed.

Another member of this family is the Black Helvella *(Helvella lacunosa)*. It's similar to white helvella, but the undulating cap is charcoal-gray tinged with lilac, and is sometimes curled up to reveal the slightly paler underside.

Collecting and Storing

Pull the mushrooms from the moss, then cut away the earthy bases before placing your find in the basket.

Helvellas will keep 2 days in the refrigerator.

Preparation and Cooking

Wash under cold running water, making sure you get rid of any tiny insects lurking in the folds of the stem. *This mushroom must not be eaten raw:* it contains toxins which can only be destroyed by cooking.

You can cook helvellas as you would morels, but they do not have the same distinctive flavor and can be rather chewy.

Since you're only likely to find helvellas in small quantities, serve them as an appetizer. Dust them with flour and fry in butter with garlic, mixed herbs, seasoning and white wine; reduce the juices by about half before serving.

White helvella (Helvella crispa)

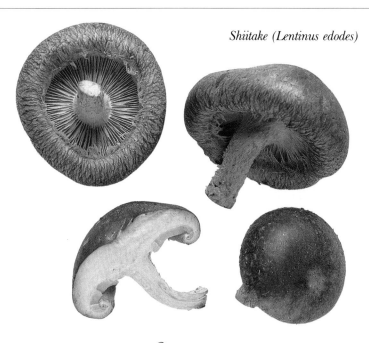

Shiitake (Lentinus edodes)

Shiitake

Shiitake Mushrooms *(Lentinus edodes)* were first grown by the Chinese, but it was the Japanese who really developed successful methods of cultivation. These mushrooms were traditionally grown on oak, teak or mahogany logs, but today's growers use hardwood sawdust contained in plastic mesh.

The fresh caps are buff-brown, with a shaggy in-rolled margin; they feel like chamois leather. Raw shiitake have a pronounced mushroomy taste with a slight peppery bite, but the flavor grows milder after cooking.

Shiitake are the most common Asian dried mushrooms sold; if you go into a Chinese market and ask for dried mushrooms, shiitake are probably what you'll get.

Buying and Storing
Fresh shiitake are becoming increasingly common in supermarkets. It's best to eat them within a day or two of purchase.

The dried caps, available in Asian markets and well-stocked supermarkets, will keep indefinitely in an airtight container. The dried mushrooms have a delicious, concentrated taste which is far more distinctive than that of fresh shiitake; you need just 4 to 6 caps to flavor a recipe. It's well worthwhile keeping a jarful in the cupboard to add to clear soups, stews and stir-fries.

Preparation and Cooking
If you are using dried shiitake, soak them in water 20 minutes to soften, then drain (you can add the soaking liquid to the dish you're making, or save it for stock). Remove the tough, leathery stems from both fresh and soaked mushrooms.

If you sauté the fresh mushrooms, they'll absorb all the fat or oil and become rather greasy. It's better to stuff them with a savory filling and bake them. You can also blanch and cool them, then toss into a green or vegetable salad.

Shiitake are delicious added to clear soups or stir-fries, or cooked with chicken; chicken and shiitake in a black bean sauce makes a wonderful meal (see page 83). You can also steam them with white fish. Cooked shiitake taste marvelous with tempura vegetables and prawns.

Wood Ear

A tree fungus, the Wood Ear *(Auricularia polytricha)* is related to Cloud Ear (see page 47)—and in fact may be sold under that name. To add to the confusion, you may also see it labelled Silver Ear, Tree Ear or Dry Black Fungus.

By any name, the gnarled clusters aren't pretty, looking something like fused-together lumps of coal. But looks aren't everything: the Chinese consider wood ears a delicacy and cultivate several varieties, ranging in size from 1 to 6 inches across. One kind is a small and fragile, brownish-black on one side and gray-brown on the other; it is contorted and looks like a crumpled leaf. A larger member—quite robust in comparison to its dainty friend—is charcoal-gray on top, with a buff underside which feels like suede.

Preparation and Cooking

You need use only a few clusters for any dish, since wood ears triple in size when soaked. To reconstitute, cover the fungus with warm water and let soak 30 minutes; as the clusters begin to soften, separate them to allow any trapped dirt to fall to the bottom of the bowl. In fact, it is advisable to change the water several times during soaking to ensure that the fungus gets clean.

The softened wood ears turn brown and become rubbery and extremely chewy; their flavor is mild. Thin slivers may be added to stir-fried or steamed dishes.

Straw Mushroom

Popular in Southern China, Southeast Asia and Madagascar, the Straw (or Padi-straw) Mushroom *(Volvariella volvacea)* is grown outdoors on rice straw. It has a small, grayish-brown cap. It is marketed fresh, dried or canned; the canned or bottled kind is sold in Chinese grocery stores and some supermarkets.

Straw mushrooms can be stir-fried with other vegetables (bean sprouts, miniature corn-on-the-cob, green onions, slivered fresh ginger for seasoning) or added to seafood, tofu or steamed chicken dishes.

Matsutake

This species grows in pine forests in Japan, where it is considered a great delicacy. Matsutake *(Tricholoma matsutake,* also known as *Armillaria matsutake)* is a large mushroom—the cap may reach 10 inches in diameter.

A number of close relatives, such as *Armillaria ponderosa* (and possibly the authentic matsutake itself), can be found in North America, particularly on the Pacific Coast during the fall. These now appear in some markets. If you cannot find matsutake, you may use blewits *(Clitocybe nuda,* see page 35) as a substitute, though they lack the matsutake's exotic fragrance and flavor.

Matsutake (Tricholoma matsutake)

*Wood ear—small variety
(Auricularia polytrica)*

*Wood ear—large variety
(Auricularia polytrica)*

Straw mushroom (Volvariella volvacea)

Cultivated Mushrooms

Until comparatively recently, mushrooms were a luxury food. But today, everyone buys them—only the potato is more popular. Furthermore, many markets are beginning to carry more kinds of mushrooms than just the familiar white buttons, a trend which can be expected to continue.

About half the recipes in this book use cultivated mushrooms. The growing process was pioneered by the French in the mid-17th century; in those days, mushrooms were grown in caves, but nowadays they're raised in specially darkened buildings.

Farmed mushrooms (the species *Agaricus bisporus,* see page 12) are grown on elaborately mixed compost in wooden trays. The trays are first pasteurized; after that, laboratory-prepared spores are sprinkled over the surface. Tiers of trays are placed in an incubating chamber until the mycelium—the threadlike root system—covers the compost. This white "cobweb" is coated with a layer of peat and chalk, and the trays are stacked in the growing houses. The first flush of mushrooms appears within 5 weeks. They are picked by hand; four or five crops are harvested from each tray.

In most markets, you'll find White mushrooms—small to medium in size, with tightly closed to slightly open caps. When a recipe calls for "button mushrooms," this is the kind to use (strictly speaking, the "button" designation refers only to fully closed mushrooms). Brown-capped varieties, including the Brown mushroom, are also available; these are reputed to have a fuller flavor than the white type. In some stores, you'll find very large cultivated mushrooms, good for stuffing. Cultivated oyster mushrooms (see page 30) are now on the market, too.

Buying and Storing

Mushrooms are sold loose and in prepacked baskets. In either case, make sure you select fresh, firm mushrooms with no blemishes.

Cultivated mushrooms are available all year at roughly the same price, so there is no need to freeze them. However, if you wish to do so, freeze them uncovered, then pack into containers; or freeze as duxelles (see page 11).

Preparation and Cooking

Do not peel cultivated mushrooms; simply trim the stems and wipe the mushrooms with a damp cloth to remove any dirt. To keep sliced raw mushrooms from discoloring, sprinkle them with lemon juice.

Small button mushrooms are good to eat raw, either tossed into salads or on their own with a dip or dressing. You might slice them, then coat with an herb-lemon vinaigrette; add avocado or bacon for a delicious luncheon salad. Packed in olive oil with spices, garlic and black peppercorns, small mushrooms will keep for several months in sealed jars; preserved this way, they make attractive gifts.

Cooked white or brown mushrooms are an ideal addition to soups such as the Crab, Corn & Mushroom Soup on page 66. They can also be added to risottos or casseroles, or made into a sauce to be eaten with pasta. If the caps are firm enough, skewer them along with the prawns in Prawn & Bacon Kabobs (see page 79).

The rich flavor of brown mushrooms makes them a good ingredient for stews. They can also be coated with a savory butter or stuffed and baked.

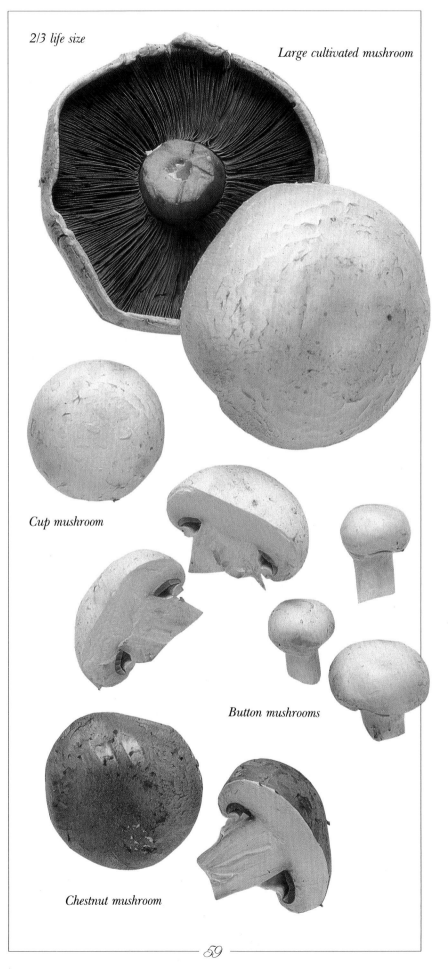

2/3 life size

Large cultivated mushroom

Cup mushroom

Button mushrooms

Chestnut mushroom

Truffles

There are three famous edible species of truffles, all highly prized for their intense flavor: the Summer or English Truffle *(Tuber aestivum)*, the Perigord or Black French Truffle *(Tuber melanosporum)* and the Piedmont or White Italian Truffle *(Tuber magnatum)*. Unfortunately, none of these subterranean tubers has been found in North America, though plenty of their underground relatives do grow in the United States and Canada. Some of the North American types are edible, and there are those who think the Oregon White Truffle *(Tuber gibbosum)* as good as the white European sort.

The truffles found in France's Perigord region are generally considered the most coveted, though the Italians would probably dispute this claim! France and Italy are rivals where truffles are concerned— despite the fact that white and black truffles differ completely in flavor and aroma, and are used in different ways. Both countries do agree on one point, though: they dismiss the English truffle because it is smaller and supposedly inferior in aroma.

The Piedmont or Italian truffle has a distinctive peppery taste and is usually eaten raw, whereas the French and English types have a marvelous fragrance which pervades all other foods they contact; they are added to savory dishes and sauces for their scent, but once cooked they have very little flavor of their own.

Truffles are an expensive luxury. Good ones are distinguished by their strong fragrance; they feel light in proportion to their size and give slightly when squeezed. They are best eaten fresh, since preserving considerably diminishes their perfume and flavor.

The truffle's legendary reputation as a powerful aphrodisiac is grossly exaggerated, though truffles do stimulate the appetite.

Truffles are supposed to favor well-drained ground beneath deciduous trees, especially oaks. They grow in small clusters 4 to 8 inches below the soil, and they do not venture beyond the tree's leafy canopy. Most Perigord truffles are found in oak plantations, where they grow spontaneously—they weren't planted, they just arrived.

The French have been trying to increase production by cultivating truffles: they have impregnated young oaks with the spores and planted mycelium at the bases of the trees, but so far these methods haven't met with success.

TRUFFLE HUNTING

It is extremely difficult—and requires endless patience—to find truffles without animals. Nowadays nobody in Britain seems to gather truffles—or if they do, they are extremely secretive about their forays. In France, the harvest is carried out in a haphazard manner; the detective work has traditionally been done by pigs, which have an insatiable appetite for truffles (this may explain why they receive a share of their precious find). Guided by the pungent odor, which apparently has some chemical similarity to the scent of a pig in season, the animals sniff out the truffles and grub them up with their snouts.

Of late, beset by difficulties in controlling their swine, the French have followed the Italians' lead and begun training truffle-hunting dogs. It's true that dogs don't have as sensitive a nose as pigs do, but they are reliable and do not tire as quickly. They must be carefully looked after to avoid "spoiling" their noses, though. Young dogs are paired with old hounds, and a truffle is rubbed on their noses to give them the scent. Once the dogs have located the truffles, their masters help dig.

In the absence of a trained pig or dog, there's another possible method of locating truffles: find a potential site, then search for animal scratchings, especially those of badgers and squirrels, or look for a cloud of flies hovering just above the ground. Both animals and flies will have been lured by the aroma of the developing tubers. By continuing the excavations of the woodland population, you may indeed uncover an underground mushroom, but don't expect to find a European truffle. Please fill in your holes and leave the area looking tidy.

Summer truffle (Tuber aestivum)

Summer Truffle

The sweet-scented, nutty-tasting Summer or English Truffle *(Tuber aestivum)*, also known as the Bath Truffle, has become a rarity. Its season runs from late summer to fall; it grows in light, dry, chalky soil and seems to favor beech trees to oaks. Blue-black and warty-skinned, it can be distinguished from its French twin by its marbled beige-white flesh. Size varies from 1-1/4 to 2-1/2 inches across.

Summer truffles were once found in the southern counties of England, especially on the Downs in Hampshire, Wiltshire and Kent. In fact, there used to be a thriving truffle trade in England—dogs were used to sniff out the tubers in beech forests. The last record of professional truffle-hunting concerns the activities of Alfred Colins who, with two dogs, scoured the woods near his home in Winterslow, in Wiltshire. After a long day's work, he carried his dogs home on his bicycle in specially made leather baskets.

Sadly, the summer truffle seems to have all but vanished, but I am sure that some still hide beneath the soil; it is not the truffles that have disappeared, but those who know how to find them.

You cannot buy summer truffles. Those pictured above were found in Gloucestershire by a friend who lent them to me to photograph. As they had to be returned the next day, I kept some in a jar of rice and some in the refrigerator next to the butter. The rice smelled and tasted wonderful in a simple risotto; I appreciated the savory butter on some hot toast and realized I was eating what might be called "poor man's truffle"—a delicious treat with which to start the day.

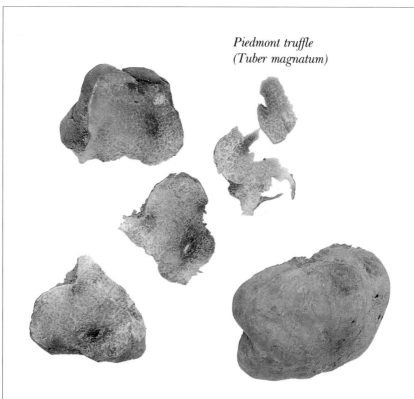

*Piedmont truffle
(Tuber magnatum)*

Piedmont Truffle

The Piedmont Truffle *(Tuber magnatum)* is also called the White Truffle— probably to distinguish it from the black tubers, because in reality it isn't white; the outside is a dirty brown, the inside beige. These truffles are found from late fall until early spring in Northern Italy—in Tuscany, Romagna and Piedmont, where the finest ones grow. The Italians use specially trained dogs to hunt them out. Piedmont truffles are usually the size of bantam eggs, but some may be as large as tennis balls.

Buying and Storing
If you cannot locate a fresh supply, try canned or bottled white truffles. Whether because of their scarcity or their delicate flavor and scent, these truffles are the most expensive variety and should be used sparingly. If you can afford a fresh white truffle, it can be used to flavor pasta: store it overnight in the refrigerator with some fresh spaghetti, or put it into a jar with dry tagliatelle.

Preparation and Cooking
Brush away the surface dirt and wipe with a damp cloth if necessary. These truffles are not usually cooked: high temperatures can spoil the flavor, and the magical aroma disappears. If you want to heat them, just warm gently with a little butter.

Scatter paper-thin slices over plain pasta dishes or simple risottos, or place on top of a cheese fondue. Like the humbler mushrooms, these truffles go well with eggs, and shavings are delicious sprinkled over scrambled eggs or omelets; the Italians grate them over fried eggs.

Add slivers to *bagna cauda* (see page 71) and serve with crudités. White truffles are also popular with chicken or turkey breasts and cheese, or as a garnish for thin veal scallops.

Perigord Truffle

A few peasant farmers in France adhere to the old traditions and go truffle-hunting with pigs, but most of the French now use trained dogs. The finest truffles are gathered from around Perigord and Magny.

The Perigord or Black Truffle *(Tuber melanosporum)* has a rough, blue-black skin that becomes brownish with age. The cut flesh turns violet-black and is covered by a fine network of whitish veins. Shape varies, as does size—the tubers may be as small as a walnut or as large as a fist. They have a penetratingly sweet aroma.

Buying and Storing
In France, fragrant black truffles are available during fall and early winter. They should feel softish to the touch and have a perfumy aroma. If you cannot find fresh ones, try the canned or bottled kind.

These expensive tubers only keep for a week, but you can easily extend their use and flavor, since their exotic odor is extremely pervasive. Add one or two black truffles to a jar of rice or put them in the refrigerator next to some butter or anything else you like—set one in the middle of a bowl of eggs, for example. The scent will penetrate the shells by the next day, and the eggs will taste delicious.

Preparation and Cooking
To clean a Perigord truffle, use a soft brush to remove all the surface dirt imbedded in the knobby skin; or wash under cold running water if necessary.

Cooking truffles whole in champagne or wine is sheer extravagance. They are a perfect flavoring for stuffings, game or duck pâté; the French add them to *foie gras.* (Canned truffles are not really suitable to use in pâtés, since their delicate flavor tends to be lost.) When you use preserved truffles, remember to add the juice.

Black truffles are delightful with eggs. Tiny pieces can be mixed in with scrambled eggs or used to season soufflés. Thin slivers are a lovely topping for eggs *en cocotte* or omelets enriched with cream and Madeira.

A traditional delicacy is chicken or turkey flavored with truffles: thin slices of truffle are placed under the skin of the uncooked bird, which is then chilled overnight (to allow the truffle flavor to penetrate the flesh) before being stuffed and roasted.

A crafty French ploy to make a little bit of truffle look like more is to mix the minced truffles with finely chopped horn of plenty (see page 20). You can use this mixture to stuff a pork tenderloin.

Perigord truffle
(Tuber melanosporum)

Cream of Mushroom Soup

3 tablespoons butter
1-1/2 pounds mushrooms (see *Note*),
   coarsely chopped
1 shallot, finely chopped
1/2 garlic clove, chopped
2 tablespoons all-purpose flour
1/2 cup white wine
3 cups stock (see *Note*)
1/2 cup whipping cream
Salt and pepper to taste

To Garnish:
1/4 cup whipping cream
Parsley sprigs

In a large saucepan, melt butter. Add mushrooms, shallot and garlic; cook a few minutes, stirring. Stir in flour and cook, stirring, 1 minute. Add wine and stock, a little at a time, stirring well after each addition. Bring to a boil; then reduce heat, cover and simmer 20 minutes.

In a food processor or blender, puree soup (in batches, if necessary); return soup to pan. Over low heat, stir in 1/2 cup cream, then add salt and pepper.

Transfer to warmed soup bowls. Swirl 1 tablespoon cream atop each bowlful; garnish with parsley sprigs.

Makes 4 servings.

Note: Button mushrooms give this soup a good color. If wild mushrooms are available, include some of them, too; they'll add a superior flavor. Fairy rings, morels, and meadow or horse mushrooms are good choices.

If possible, use mushroom stock for this soup; otherwise, use vegetable or chicken stock.

Carrot Soup with Croûtons

2 tablespoons butter
1 small onion, finely chopped
1 pound carrots, sliced
1 (2-inch) piece fresh gingerroot,
  chopped
3-3/4 cups chicken stock
1 bay leaf
Pinch of freshly grated nutmeg
1/3 cup whipping cream
3 tablespoons brandy
Salt and pepper to taste

Mushroom Croûtons:
1/4 cup butter, room temperature
3 ounces mushrooms, finely chopped
2 teaspoons finely chopped parsley
1/2 small garlic clove, pressed or
  minced
8 slices French bread

To Garnish:
Cilantro or parsley sprigs

In a large saucepan, melt butter; add onion and sauté about 5 minutes or until softened and golden. Add carrots and ginger and cook 2 to 3 minutes. Pour in stock and bring to a boil; add bay leaf and nutmeg. Reduce heat, cover and simmer 20 to 25 min-

utes or until carrots are soft. Remove bay leaf.

In a food processor or blender, puree soup (in batches, if necessary); return soup to pan. Over low heat, stir in cream, brandy, salt and pepper.

To make croûtons, mix room-temperature butter, mushrooms, parsley and garlic in a small bowl. Toast bread on one side only; turn slices over and spread other sides with mushroom butter. Just before serving, toast slices under the broiler until hot and golden.

Serve soup in warmed wide soup bowls; float 2 croûtons on each serving and garnish with cilantro or parsley sprigs.

Makes 4 servings.

Note: The soup may be prepared in advance and reheated; the croûtons may be spread with mushroom butter, then toasted at the last minute.

Crab, Corn & Mushroom Soup

3-3/4 cups chicken stock
1 (1-inch) piece fresh gingerroot,
    coarsely chopped
2 green onions, coarsely chopped
About 1 cup cooked corn kernels
3 tablespoons peanut oil
4 ounces button mushrooms,
    quartered
6 ounces lump crabmeat, flaked into
    small pieces
2 eggs, lightly beaten
Salt and pepper to taste

To Garnish:
Fresh cilantro leaves

In a large saucepan, combine stock, ginger and green onions. Bring to a boil; then reduce heat, cover and simmer 15 minutes.

Meanwhile, in a food processor or blender, smoothly puree half the corn kernels.

Strain stock; discard ginger and onions.

In same pan, heat oil; add mushrooms and sauté 2 minutes or until softened. Stir in remaining corn kernels, corn puree and strained stock. Bring to a boil; then reduce heat and simmer, uncovered, 10 minutes.

Just before serving, stir in crabmeat and bring to a gentle simmer. Pour in beaten eggs in a slow stream, stirring constantly; add salt and pepper.

Serve in warmed soup bowls, garnished with cilantro.

Makes 4 servings.

Note: This soup may be prepared in advance up to the point when the crab and eggs are added.

Chicken Soup, Chinese Style

5 dried shiitake mushrooms
3 cups chicken stock (use strong
   stock)
2 slices plus 1/4 teaspoon finely
   chopped fresh gingerroot
2 cilantro sprigs
1 (4-inch) piece lemon grass
Pinch of sugar
2 green onions, shredded
2 teaspoons soy sauce
2 tablespoons dry sherry or rice wine
4 ounces cooked chicken, cut or torn
   into small strips
Szechuan pepper to taste (see *Note*)

Place mushrooms in a small bowl; cover with warm water and let soak 20 minutes. Drain, reserving soaking liquid. Rinse mushrooms well; trim off tough stems and slice caps. Set aside.

In a large saucepan, combine stock, mushroom liquid, ginger slices, cilantro sprigs, lemon grass and sugar. Bring to a boil; then reduce heat, cover and simmer 30 minutes. Strain through a fine sieve; discard ginger, cilantro and lemon grass. Return stock to pan and add mushrooms, chopped ginger and remaining ingredients. Bring to a boil; then reduce heat, cover and simmer 10 minutes. Adjust seasoning. Serve piping hot, in warmed soup bowls.

Makes 4 servings.

Note: You can buy Szechuan peppercorns—tiny, pungent-tasting, reddish-brown berries—in Asian markets and well-stocked supermarkets. Before using the peppercorns, bring out their flavor by toasting them in a dry frying pan over medium-high heat 5 minutes. Then grind, using a mortar and pestle.

Potato & Mushroom Soup

3 tablespoons butter
1 onion, finely chopped
1-1/2 pounds potatoes, peeled, cut
   into 1/2-inch cubes
2 bacon slices, diced
12 ounces brown or chestnut
   mushrooms, quartered
3-3/4 cups beef or chicken stock
1-1/4 cups whole or extra-rich milk
Sprinkling of freshly grated nutmeg
Salt and pepper to taste

To Garnish:
Crumbled crisp-cooked bacon
Snipped chives

In a large saucepan, melt butter; add onion and sauté about 5 minutes or until pale golden. Stir in potatoes, diced bacon and mushrooms; cook 1 minute. Add stock, milk and nutmeg. Bring to a boil; reduce heat, cover and simmer 20 minutes or until potatoes are soft, stirring occasionally.

In a food processor or blender, puree half the soup; then return to remaining soup in pan. Season with salt and pepper.

Pour into warmed soup bowls; garnish with crumbled bacon and chives.

Makes 6 servings.

Smoked Salmon Pâté

**9 ounces smoked salmon, thinly
  sliced
5 tablespoons butter
9 ounces button mushrooms, sliced
About 1/3 cup yogurt or dairy sour
  cream
1-1/2 teaspoons lemon juice
2 teaspoons chopped fresh chervil
Salt and pepper to taste**

To Garnish:
**Chervil sprigs
Lemon slices**

To Serve:
Melba toast or other crisp crackers

Using two-thirds of the salmon slices,
line bottoms and sides of 4 (1/2-cup)
ramekins. Set aside. Coarsely chop
remaining salmon.

In a large saucepan, melt butter.
Add mushrooms; sauté about 5 minutes or until soft. Transfer to a food
processor or blender and let cool;
then add chopped salmon and puree
until mixture is smooth. Transfer to a
bowl and fold in yogurt or sour
cream, lemon juice, chopped chervil,
salt and pepper.

Divide pâté among prepared ramekins, folding any overhanging salmon slices over top. Cover each rame-
kin with plastic wrap and chill at least
1 hour.

To serve, turn out salmon-
wrapped pâté onto serving plates and
garnish with chervil sprigs and lemon
slices. Serve with melba toast.

Makes 4 servings.

Oyster Mushroom Beignets

Vegetable oil for deep-frying
1 pound oyster mushrooms

Tangy Dip:
2/3 cup dairy sour cream
1 tablespoon lemon juice
1/2 garlic clove, pressed or minced
1 teaspoon honey
1/4 cup finely chopped fresh lemon balm or mint
Salt and pepper to taste

Beer Batter:
1 cup all-purpose flour
1/2 teaspoon salt
2 eggs, separated
2/3 cup beer
2 tablespoons olive oil

To Garnish:
Lemon balm or mint sprigs

Prepare dip by mixing all dip ingredients in a small bowl; set aside. To prepare batter, sift flour and salt into a bowl; add egg yolks, then gradually whisk in beer to form a smooth batter. Stir in oil. In a clean bowl, whisk egg whites until they hold soft peaks; then fold into batter.

In a deep, heavy saucepan, heat 3 or 4 inches of oil to 350F (175C). Dip mushrooms into batter to coat; lower into hot oil, a few at a time (do not crowd pan). Cook 2 to 3 minutes or until crisp and golden. Drain on paper towels and keep hot while cooking remaining mushrooms.

Spoon dip into a small serving bowl; garnish with lemon balm or mint sprigs. Serve beignets with dip.

Makes 4 servings.

Variation: Substitute button mushrooms or chanterelles in place of oyster mushrooms.

Bagna Cauda

1-1/4 cups whipping cream
3 garlic cloves, pressed or minced
2 (2-oz.) cans anchovy fillets, drained, chopped
1/4 cup unsalted butter, cut into pieces

To Garnish:
Parsley sprigs

To Serve:
Whole button mushrooms
Radishes
Cubes of crusty bread

In a small saucepan, combine cream, garlic and anchovy fillets. Bring to a boil; then reduce heat and simmer gently, uncovered, 12 to 15 minutes or until smooth and thickened. Add butter; stir until melted.

Transfer to a serving dish and garnish with parsley sprigs. Serve with mushrooms, radishes and bread cubes for dipping.

Makes 4 servings.

Note: This dish comes from the Piedmont region of Northern Italy, where it was traditionally served as a dip for cardoon, a locally grown edible thistle. I think it is superb with firm white button mushrooms. Keep the dip hot at the table over a candle or alcohol flame.

Marinated Mushrooms

6 tablespoons virgin olive oil
1 teaspoon coriander seeds, lightly
   crushed
1/2 teaspoon cracked bay leaves
1 pound whole large cultivated or
   button mushrooms
Finely grated peel of 1/2 lime
Juice of 1 lime
1 garlic clove, pressed or minced
2 tablespoons white wine or cider
   vinegar
Pinch of sugar
Salt and pepper to taste

To Garnish:
1 tablespoon chopped parsley
Lime slices
Parsley sprigs

To Serve:
Crusty French bread

In a large saucepan, heat 3 tablespoons oil. Add coriander seeds and cook 1 minute. Stir in bay leaves and mushrooms; cook over low heat 5 to 7 minutes or until mushrooms are just tender. Remove from heat and add lime peel, lime juice, garlic, vinegar, sugar, salt and pepper. Mix well, then transfer to a shallow bowl and let cool. Cover and chill at least 2 hours.

Adjust seasoning. Garnish mushrooms with chopped parsley, lime slices and parsley sprigs. Serve with crusty French bread.

Makes 4 servings.

Variation: To give the dish a hint of anise flavor, substitute fennel seeds for coriander seeds.

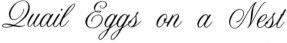

Quail Eggs on a Nest

Chicory leaves
1/4 cup walnut oil
1/2 garlic clove, pressed or minced
4 very large (about 4-inch-diameter)
   cultivated mushrooms
12 uncooked quail eggs (available at
   some gourmet markets)
3 bacon slices, cut into short, thin
   strips
1 tablespoon plus 2 teaspoons white
   wine vinegar
Salt and pepper to taste

To Garnish:
2 teaspoons snipped chives
Whole chives

Line 4 individual serving plates with chicory; set aside. In a large nonstick frying pan, heat 2 tablespoons oil; add garlic and mushrooms. Cook over low heat about 5 minutes or until mushrooms are just tender. Place one mushroom, stalk side up, atop chicory on each plate.

Add remaining 2 tablespoons oil to pan. When oil is hot, add quail eggs, in 4 batches; cook each batch about 30 seconds. Carefully arrange eggs over mushrooms.

Add bacon to pan and cook over medium-high heat until crisp. Deglaze pan with vinegar, stirring to scrape up browned bits; season with salt and pepper.

Pour hot dressing equally over salads and sprinkle with snipped chives; add a few whole chives alongside each mushroom "nest." Serve immediately.

Makes 4 servings.

Morels & Oysters in Brioches

4 small brioches (available at many
   bakeries)
6 ounces fresh morels
1/4 cup butter
1/2 small onion, finely chopped
1/4 cup dry white wine
3/4 cup whipping cream
6 freshly shucked oysters, halved
1 teaspoon cornstarch blended with 2
   teaspoons water
2 to 3 teaspoons lemon juice
Salt and pepper to taste

To Garnish:
Assorted salad greens

Preheat oven to 325F (160C).

Cut tops from brioches; carefully
hollow out each brioche (you can reserve the centers to make bread
crumbs). Place brioches and tops on a
baking sheet and set aside.

To prepare morels, slice in half
lengthwise and place in a bowl of
salted water. Let soak 15 minutes,
then rinse carefully. Blanch in boiling
water 2 minutes; drain.

In a medium-size saucepan, melt
butter; add onion and sauté 2 minutes or until softened. Stir in wine and
mushrooms and cook, stirring, over
high heat until almost all liquid has
evaporated. Add cream and oysters.
Reduce heat; cook until edges of oysters just begin to curl and cream is
almost boiling. Stir in cornstarch mixture; cook, stirring, until thickened.
Add lemon juice, salt and pepper.

Divide mixture among brioches.
Set brioche tops in place and bake 15
to 20 minutes or until hot. Serve immediately, garnished with salad
greens.

Makes 4 servings.

Variation: Cultivated mushrooms
can be used in place of morels. Cut
into quarters or slices before cooking.

Chèvre & Mushroom Croustades

5 ounces wild mushrooms (see *Note*)
1 to 2 teaspoons sesame oil
2 tablespoons butter
2 cups fresh white bread crumbs
1 tablespoon plus 1 to 2 teaspoons
   sesame seeds
6 ounces soft chèvre (goat cheese)
1 tablespoon finely chopped
   sun-dried tomatoes (see *Note*)
1 tablespoon chopped fresh basil
1 tablespoon virgin olive oil

To Garnish:
Fresh basil leaves
Assorted salad greens

Preheat oven to 400F (205C). Chop mushrooms coarsely unless they are very small; set aside. Brush 4 (3- to 4-inch) tartlet pans with sesame oil and set aside.

In a small saucepan, melt butter; stir in crumbs and 1 tablespoon sesame seeds. Divide crumb mixture evenly among tartlet pans, pressing it firmly against pan bottoms and sides. Bake 12 to 15 minutes or until pale golden and crisp.

Meanwhile, in a small bowl, mix chèvre, sun-dried tomatoes and

chopped basil; set aside.

In a small frying pan, heat olive oil; add mushrooms and sauté 1 minute. Reserve a few mushrooms for garnish, then divide the rest equally among baked croustade shells. Top with chèvre mixture, spreading it evenly to fill shells. Top with reserved mushrooms and sprinkle with remaining 1 to 2 teaspoons sesame seeds. Return to oven; bake 10 minutes or until filling is hot.

Serve hot or warm, garnished with basil sprigs and salad greens.

Makes 4 servings.

Note: Use chanterelles, fairy rings, horse mushrooms, cepes (porcini), oyster mushrooms or a mixture of wild mushrooms; or use sliced button mushrooms.

Sun-dried tomatoes are available in many delicatessens and gourmet food shops. They're sold in jars, preserved in olive oil, with or without salt. For this recipe, unsalted tomatoes are best.

Ravioli with Three Cheeses

12 flat-leaf parsley leaves
2 tablespoons butter, melted
1 to 2 tablespoons freshly grated
  Parmesan cheese

Egg Pasta:
1 cup all-purpose flour
1 egg
1 teaspoon olive oil
Pinch of salt

Cheese Filling:
1 ounce dried cepes (porcini)
6 tablespoons ricotta cheese
3 ounces smoked mozzarella or
  provolone cheese, finely chopped
2 tablespoons whipping cream
Pepper to taste

To make pasta, sift flour onto a work surface. Make a well in center; add egg, oil and salt and beat with a fork to blend. Gradually work flour into egg mixture; then knead together, adding a little water if dough seems dry. On lightly floured work surface, knead dough at least 5 minutes or until smooth and elastic. Wrap in plastic wrap and set aside 1 hour.

To make filling, place cepes in a small bowl; cover with warm water and let soak 20 minutes. Drain; rinse well to remove any grit. Chop coarsely, place in a clean small bowl, and mix in cheeses, cream and pepper.

On a lightly floured surface, roll out dough paper-thin; cut into 24 squares. Divide filling among 12 squares and place a parsley leaf on each; cover with remaining pasta squares. Parsley should be visible through pasta. Press edges of each ravioli to seal; trim edges, then crimp with a fork. Let stand 10 minutes to dry slightly before cooking.

Cook in a large saucepan of boiling salted water 3 minutes; drain well. Allow 3 ravioli per person; drizzle with butter and sprinkle with Parmesan cheese.

Makes 4 servings.

Gruyère & Mushroom Tart

Egg Pastry:
1-1/2 cups all-purpose flour
Pinch of salt
1/2 cup butter, cut into pieces
1 egg yolk
2 to 3 tablespoons water

Tart Filling:
1 tablespoon butter
1 shallot, finely chopped
6 ounces mushrooms (see *Note*),
    chopped
4 ounces Gruyère cheese, grated
3 eggs, lightly beaten
1 cup whipping cream
Pepper to taste
Generous pinch of freshly grated
    nutmeg

To Serve:
Tossed green salad

To make pastry, sift flour and salt into a large bowl; cut in butter until mixture resembles fine crumbs. Add egg yolk and 2 to 3 tablespoons water; mix to form a firm but pliable dough. Roll out dough on a lightly floured surface and use to line an 8-inch tart pan; prick shell lightly and chill 15 minutes. Meanwhile, preheat oven to 425F (220C). Bake tart shell blind 15 minutes. Remove from oven; reduce temperature to 375F (190C).

Meanwhile, prepare filling. In a small saucepan, melt butter. Add shallot and sauté 4 to 5 minutes or until golden. Add mushrooms and cook 2 to 3 minutes or until tender. Spread mixture in prepared pastry shell.

In a bowl, mix cheese, eggs and cream; season with pepper. Pour into pastry shell and sprinkle with nutmeg. Bake about 35 minutes or until filling is just set. Serve warm, cut into wedges, with salad.

Makes 6 servings.

Note: Use any mushroom variety for this recipe, or combine several kinds. Dried mushrooms give a good flavor; try morels or cepes (porcini). You will need about 1 ounce; reconstitute as directed on page 100.

Mussel-stuffed Crêpes

2/3 cup dry white wine
1 garlic clove, pressed or minced
2 shallots, finely chopped
2 pounds mussels, scrubbed
2 tablespoons butter
6 ounces wild mushrooms (see *Note*), chopped
6 tablespoons plain yogurt
2 teaspoons chopped parsley
8 crêpes (use your favorite recipe)

Tomato Sauce:
1 tablespoon olive oil
1 small onion, chopped
1 pound tomatoes, chopped
Pinch of sugar
2 teaspoons chopped parsley
Salt and pepper to taste

Preheat oven to 375F (190C).

In a large saucepan, combine wine, garlic, half the shallots and 1/3 cup water. Add mussels; cover and cook over high heat 3 to 5 minutes or until opened. Discard any unopened mussels. Drain, reserving cooking liquid. Remove mussels from shells (leave a few unshelled for garnish). Set aside.

Prepare Tomato Sauce: In a saucepan, heat oil. Add onion and sauté over low heat 4 to 5 minutes. Add tomatoes, mussel liquid and sugar; cover and cook 5 minutes. Remove from heat; add parsley. Puree in a food processor or blender; sieve, return to pan and cook until thick. Add salt and pepper.

Melt butter in a frying pan, add remaining shallots and sauté 2 minutes. Stir in mushrooms and cook 2 to 3 minutes or until tender. Remove from heat. Stir in mussels, yogurt and parsley; adjust seasoning.

Divide filling among crêpes; fold crêpes into quarters. Arrange in a single layer in a buttered baking dish; bake, uncovered, 15 minutes. Garnish crêpes with unshelled mussels; serve with heated Tomato Sauce.

Makes 4 servings.

Note: Use parasols, chanterelles, morels, fairy rings or cepes (porcini).

Prawn & Bacon Kabobs

12 cooked large prawns, shelled,
   deveined, halved crosswise
12 bacon slices, halved
Walnut oil

Mushroom & Rice Balls:
1 tablespoon butter
1 tablespoon walnut oil
1 shallot, finely chopped
1 garlic clove, pressed or minced
4 ounces button mushrooms, finely
   chopped
About 2 cups cooked long-grain rice
1-1/4 cups fresh white bread crumbs
1 tablespoon plus 1 teaspoon chopped
   fresh cilantro
1 egg, beaten
Salt and pepper to taste

To Garnish:
Assorted salad greens

First make Mushroom & Rice Balls:

In a saucepan, melt butter in oil. Add shallot and garlic and cook 1 minute or until softened. Stir in mushrooms, rice, crumbs, cilantro and egg; season with salt and pepper. Preheat broiler. Shape rice mixture into 16 walnut-size balls; place on a greased baking sheet. Broil about 3 minutes or until firm, turning once.

Wrap each prawn piece in half a bacon slice. Thread bacon-wrapped prawns alternately with Mushroom & Rice Balls on 8 short skewers. Place on a baking sheet; brush with oil. Broil 5 minutes or until bacon is crisp and browned, turning occasionally.

Garnish kabobs with salad greens.

Makes 4 servings.

Lemon Sole with Mushrooms

8 small skinless lemon sole fillets
1-1/4 cups fish or chicken stock
1-1/4 cups dry white wine
2 teaspoons cornstarch blended with
   1 teaspoon water
1/4 cup whipping cream
1 tablespoon chopped fresh fennel
   leaves
Salt and pepper to taste

Mushroom Stuffing:
1 tablespoon butter
1/4 cup finely chopped fennel bulb
8 ounces oyster mushrooms, chopped
1/3 cup pine nuts, toasted, chopped
1 cup fresh white bread crumbs
Juice and grated peel of 1/2 lemon

To Garnish:
Few oyster mushrooms, cooked
Fresh fennel leaves
2 teaspoons pine nuts, toasted

Preheat oven to 375F (190C).

Prepare stuffing: In a large sauce-pan, melt butter. Add chopped fennel bulb and sauté 1 minute or until softened. Add mushrooms; cook 2 to 3 minutes or until tender. Stir in pine nuts, crumbs, lemon juice and lemon peel.

Lay fish fillets flat, skinned side up. Top evenly with stuffing; roll up and place, seam side down, in a shallow baking dish. Mix stock and wine; pour over fish. Cover loosely with foil; bake 20 to 25 minutes or just until fish is opaque throughout. Carefully transfer fish rolls to a plate; keep warm. Strain cooking juices into a small saucepan, bring to a boil and reduce by half. Stir in cornstarch mixture; cook, stirring, until thickened. Remove from heat and stir in cream, chopped fennel leaves, salt and pepper.

Arrange fish rolls on warmed serving plates; pour sauce over and around them. Garnish with whole mushrooms, fennel leaves and pine nuts. Serve with wild rice.

Makes 4 servings.

Salmon with Mushrooms & Dill

4 salmon steaks
1 tablespoon lemon juice
Salt and pepper to taste

Mushroom & Dill Butter:
5 ounces small mushrooms (see *Note*)
6 tablespoons unsalted butter, room
   temperature
1/2 teaspoon grated lemon peel
2 tablespoons chopped fresh dill

To Garnish:
Blanched snow peas
Dill sprigs

Preheat oven to 375F (190C).

Place each salmon steak on a square of foil; sprinkle with lemon juice, salt and pepper. Wrap foil around salmon to enclose, then place packets on a baking sheet. Bake 10 to 15 minutes or just until fish is opaque throughout.

Meanwhile, prepare Mushroom & Dill Butter. Leave a few mushrooms whole for garnish; finely chop the rest. In a small saucepan, melt 2 table-spoons butter; add whole mushrooms and sauté 1 minute or until just tender. Remove from pan and set aside. Add chopped mushrooms to pan and cook 2 to 3 minutes or until softened. Transfer to a small bowl and let cool 1 to 2 minutes; stir in lemon peel, chopped dill and remaining 1/4 cup butter.

Remove skin and central bone from each salmon steak. Transfer salmon to warmed serving plates and top with Mushroom & Dill Butter. Garnish with snow peas, reserved whole mushrooms and dill sprigs. Serve immediately.

Makes 4 servings.

Note: Because of their small size, fairy rings and tiny button mushrooms are prettiest for this dish. You may also use a mixture of wild or cultivated mushrooms; if they are large, slice a few for garnish.

Spiced Roast Chicken

1 (3-1/2-lb.) chicken
1 tablespoon butter
2/3 cup marsala

Mushroom Stuffing:
3 tablespoons butter
1 onion, finely chopped
1 teaspoon garam masala
4 ounces button, brown or chestnut
  mushrooms, chopped
1 cup coarsely grated parsnips
1 cup coarsely grated carrots
1/4 cup minced walnuts
2 teaspoons chopped fresh thyme
1 cup fresh white bread crumbs
1 egg, beaten
Salt and pepper to taste

To Garnish:
Thyme and watercress sprigs

To Serve:
Seasonal vegetables

Preheat oven to 375F (190C).

Prepare stuffing: In a large sauce-pan, melt butter; add onion and sauté 2 minutes or until softened. Stir in garam masala and cook 1 minute.

Add mushrooms, parsnips and carrots; cook, stirring, 5 minutes. Remove from heat; stir in remaining stuffing ingredients.

Stuff and truss chicken. Place, breast down, in a roasting pan; add 1/4 cup water. Roast 45 minutes; turn chicken breast up and dot with butter. Roast about 45 more minutes or until a meat thermometer inserted in thickest part of thigh (not touching bone) registers 185F (85C). Transfer to a platter; keep warm.

Pour off and discard fat from roasting pan; add marsala to remaining cooking juices, stirring to scrape up any browned bits. Boil over high heat 1 minute to reduce slightly; adjust seasoning.

Carve chicken and garnish with thyme and watercress sprigs. Serve with stuffing, flavored meat juices and seasonal vegetables.

Makes 4 servings.

Chicken in Black Bean Sauce

12 ounces skinned, boned chicken
  breast, cut into strips
About 2/3 cup chicken stock
1 tablespoon plus 1 teaspoon sesame
  oil
1 tablespoon plus 1 teaspoon corn or
  peanut oil
4 green onions, sliced diagonally
6 ounces fresh shiitake mushrooms
  (see *Note*), sliced
8 ounces broccoli flowerets
1/4 cup preserved black beans, rinsed
2 teaspoons cornstarch blended with
  1 teaspoon water

Soy Marinade:
2/3 cup dry sherry
2 tablespoons soy sauce
1 teaspoon light brown sugar
1 garlic clove, pressed or minced
1 (1-inch) piece fresh gingerroot,
  grated
1 fresh red chile, seeded, thinly sliced

To Serve:
Hot rice or thin noodles

Prepare marinade: Combine all marinade ingredients in a shallow bowl and mix well. Stir in chicken, cover and refrigerate at least 1 hour. Lift from bowl with a slotted spoon; measure marinade and add enough stock to make 1-1/4 cups. Set aside.

In a wok or large frying pan, heat sesame and corn or peanut oils over high heat. Add chicken and stir-fry about 4 minutes or just until browned. Add green onions, mushrooms, broccoli and black beans; stir-fry about 5 minutes or just until broccoli is tender-crisp.

Mix stock-marinade mixture with cornstarch mixture. Pour into wok or pan and cook, stirring constantly, until sauce is thickened. Serve immediately, with rice or noodles.

Makes 4 servings.

Note: If fresh shiitake are unavailable, substitute button mushrooms. Or use 6 dried shiitake, reconstituted as directed on page 67.

Duck with Raspberry Sauce

4 boneless duck breasts, halved
1 tablespoon honey
1 garlic clove, pressed or minced
1/2 cup Madeira or sweet sherry
2 tablespoons butter
4 very large (about 4-inch-diameter)
** cultivated mushrooms**
4 ounces raspberries
3 tablespoons whipping cream
Salt and pepper to taste

To Garnish:
Chervil or cilantro sprigs

Using a fork, prick skin of each duck-breast half several times. Spread duck pieces with honey and garlic; place in a shallow bowl. Pour Madeira or sherry over duck, cover and refrigerate at least 1 hour.

Remove duck from bowl with a slotted spoon, reserving marinade. In a large frying pan, melt butter over high heat; add duck and cook 2 minutes or until browned, turning once.

Reduce heat and cook 10 to 12 more minutes or until skin is a deep, rich brown and meat in thickest part is firm, but still pink. Transfer to a plate and keep warm.

Increase heat and add mushrooms to pan, turning quickly in duck juices. Pour in reserved marinade and cook about 4 minutes or until mushrooms are tender; transfer mushrooms to plate and keep warm.

Add raspberries to pan and cook over high heat until liquid in pan is slightly syrupy. Remove from heat; stir in cream, salt and pepper.

To serve, slice each duck breast; arrange each on a warmed serving plate with one mushroom. Spoon sauce over and around mushrooms. Garnish with chervil or cilantro sprigs.

Makes 4 servings.

Lamb & Mushroom Blanquette

1/4 cup butter
1-1/2 pounds lean shoulder of lamb,
    cut into 1-1/2-inch pieces
2-1/2 cups lamb or chicken stock
1/2 cup dry white wine
1 onion, quartered
1/2 teaspoon saffron threads
2 egg yolks
2/3 cup whipping cream
2 teaspoons lemon juice
Salt and white pepper to taste
1 garlic clove, halved
6 ounces wild mushrooms (see *Note*)

To Garnish:
Small bundles of lightly cooked
    spring vegetables

In a large saucepan, melt 2 table-spoons butter. Add lamb and sauté 2 minutes, just to seal; do not brown. Pour in stock and wine; bring to a boil, then skim surface. Add onion and saffron; reduce heat, cover and simmer 1-1/2 to 2 hours or until lamb is tender, skimming occasionally.

Strain stock through a fine sieve.

Discard onion; set lamb aside and keep warm. Return stock to pan and boil over high heat until reduced by half. Reduce heat to low. Beat together egg yolks and cream, then whisk into stock; heat gently, but do not boil. Season with lemon juice, salt and white pepper. Keep warm.

In a small frying pan, melt remaining 2 tablespoons butter. Add garlic and mushrooms and sauté 2 to 3 minutes or until mushrooms are tender. Discard garlic.

To serve, arrange lamb on a warmed platter or individual plates. Spoon sauce over meat; arrange mushrooms alongside meat. Garnish with spring vegetables.

Makes 4 servings.

Note: Use whatever wild mushrooms are available; fairy rings, morels, oyster mushrooms and chanterelles are pretty and flavorful. Or use a combination of cultivated types.

Beef Casseroled in Stout

1/4 cup butter
1/4 cup olive oil
1-1/2 pounds chuck or round steak,
   cut into 1-inch cubes
2 tablespoons all-purpose flour
1-1/4 cups stout
2 garlic cloves, halved
1 tablespoon chopped fresh rosemary
12 pearl onions, peeled
5 ounces baby carrots
Beef stock or water, if needed
6 ounces small button mushrooms
Salt and pepper to taste

To Garnish:
Rosemary sprigs

To Serve:
New potatoes or hot rice
Crusty bread

Preheat oven to 325F (165C).

In a large saucepan, melt 2 table-spoons butter in 2 tablespoons oil. Add beef (in batches, if necessary); cook over high heat just to seal. Transfer to a casserole.

Add remaining 2 tablespoons oil to pan; stir in flour and cook, stirring, 2 minutes. Remove from heat; gradually stir in stout. Heat gently, stirring, to form a smooth sauce. Add to casserole with garlic and chopped rosemary. Cover and bake 1-1/2 hours, stirring occasionally.

In a small frying pan, melt remaining 2 tablespoons butter. Add onions and sauté 2 to 3 minutes or until browned. Stir into casserole with carrots. Check level of liquid; if needed, add a little stock or water to cover meat. Return to oven and bake 30 more minutes.

Stir in mushrooms; season with salt and pepper. Return to oven; bake 30 more minutes or until meat is very tender.

To serve, garnish with rosemary sprigs and accompany with potatoes or rice and crusty bread.

Makes 4 servings.

Filets de Boeuf en Croûtes

1/4 cup butter
4 tenderloin steaks, each about 1 inch
   thick
6 ounces fresh wild mushrooms (see
   Note), sliced if large
2 tablespoons brandy
12 ounces puff pastry
6 ounces goose liver pâté with truffles
   (see *Note*)
Salt and pepper to taste
1 egg, beaten

Preheat oven to 425F (220C).

In a large frying pan, melt butter. Add steaks and cook 1 minute on each side, just to seal. Transfer to a plate; let cool.

Add mushrooms to pan and sauté 1 minute. Pour in brandy; stir to scrape up browned bits in pan. Ignite; when flames subside, set aside.

On a lightly floured surface, roll out pastry to a 12-inch square; cut into 4 (3-inch) squares. Place a fourth of the pâté in center of each square.

Cover pâté with mushrooms and their cooking juices, then place steaks on top. Season with salt and pepper.

Brush pastry edges with water. Draw up all 4 corners of each square to center; twist to seal, then press seams together. Turn packets over, seam side down; with a sharp knife, mark top of each in a lattice pattern (don't cut all the way through). Place on a baking sheet; brush with egg. Bake 25 to 30 minutes or until pastry is well risen and golden brown.

Serve with new potatoes and a green vegetable, such as asparagus.

Makes 4 servings.

Note: Use parasols, chanterelles, horse mushrooms, or cepes (porcini) or other boletes for this recipe.

Goose liver pâté with truffles is available in cans or jars in gourmet shops and good delicatessens. Other pâtés can be substituted.

Pork with Oyster Mushrooms

1 pound pork tenderloin
3-2/3 cups white bread crumbs (from
   day-old bread)
Grated peel of 1/2 lemon
Salt and pepper to taste
1 egg, beaten with 2 tablespoons
   water
1/4 cup butter
6 ounces oyster mushrooms,
   quartered if large
1 garlic clove, pressed or minced
1/4 cup Madeira
2/3 cup whipping cream

To Garnish:
Lemon slices
Parsley sprigs

To Serve:
Green vegetables or tossed green
   salad

Cut pork tenderloin into 1/2-inch
slanting slices; pound with a flat-
surfaced mallet to flatten. Mix
crumbs, lemon peel, salt and pepper.
Dip pork pieces into beaten egg to
coat, then roll in seasoned crumbs to
coat completely.

In a large frying pan, melt 3 tablespoons butter. Add pork, several
slices at a time; sauté slices 2 minutes
on each side or until meat is no longer
pink in center. Transfer to a warmed
dish and keep warm. Wipe out pan
with paper towels.

Melt remaining 1 tablespoon butter in same pan. Add mushrooms and
garlic and sauté 2 to 3 minutes or
until tender. Remove mushrooms
from pan; keep warm. Add Madeira
to pan and boil over high heat until
liquid is reduced by half; reduce heat,
stir in cream and heat through. Season with salt and pepper.

Arrange pork and mushrooms on
warmed serving plates. Spoon sauce
over mushrooms and garnish with
lemon slices and parsley sprigs. Serve
with green vegetables or a salad.

Makes 4 servings.

Pork Tenderloin with Truffles

4 ounces fresh or 3/4 ounce dried
   horn of plenty mushrooms
2 (12-oz.) pork tenderloins
2 garlic cloves, cut into slivers
1/2 ounce black truffles, cut into
   slivers
5 tablespoons butter
Pepper to taste
1 cup dry white wine
1 teaspoon all-purpose flour
Salt to taste

To Garnish:
Thyme or chervil sprigs

Preheat oven to 375F (190C). If using dried mushrooms, place them in a small bowl, cover with warm water and let soak 20 minutes. Drain; rinse well. Chop soaked or fresh mushrooms; set aside.

Trim pork tenderloins to the same length. Slit each lengthwise down center, being careful not to cut all the way through. Lay pork out flat; make 2 more lengthwise cuts down each piece to flatten meat further.

Arrange garlic slivers over one tenderloin; cover with mushrooms and truffles. Dot with 2 tablespoons but-ter and sprinkle with pepper. Cover with second tenderloin; securely tie pieces together with string at 1-inch intervals.

In a large frying pan, melt 2 table-spoons butter; add pork and brown on all sides. Transfer pork and its cooking juices to a small roasting pan. Pour in wine; roast 20 to 30 minutes or until meat is no longer pink in center. Transfer meat to a plate; keep warm. Boil cooking juices over high heat 1 minute to reduce slightly. Mix remaining 1 tablespoon butter and flour; add to juices and cook, stirring, until thickened. Season with salt and pepper.

Cut tenderloin into slices; spoon sauce around meat and garnish with thyme or chervil sprigs.

Makes 4 servings.

Note: Black horn of plenty mush-rooms are used here to extend a tiny quantity of the rarer black truffle.

Ham & Mushroom Pasta

3/4 cup dried cepes (porcini)
2 tablespoons butter
1 shallot, finely chopped
2 ounces small button mushrooms,
quartered
2 (3-oz.) packages cream cheese with
chives, cubed
2/3 cup whipping cream
1-1/2 ounces prosciutto, cut into
strips
Pepper to taste
1 pound fresh tagliatelle

To Serve:
Tossed green salad
Cherry tomatoes

Place cepes in a small bowl; cover with warm water and let soak 20 minutes. Drain; rinse well to remove any grit.

Chop coarsely.

In a saucepan, melt butter; add shallot and sauté 2 minutes or until softened. Add cepes and button mushrooms and sauté 3 minutes or until button mushrooms are just beginning to brown. Stir in cream cheese, cream and prosciutto and cook over low heat, stirring constantly, until hot and well blended. Season with pepper.

Cook tagliatelle in plenty of boiling salted water about 3 minutes or just until al dente; drain thoroughly.

Add tagliatelle to sauce and toss well. Serve with a green salad and cherry tomatoes.

Makes 4 servings.

Risotto con Funghi

1 ounce dried cepes (porcini)
1/2 cup butter
1 small onion, finely chopped
2 cups short-grain rice, such as pearl
4 ounces fresh mushrooms (see *Note*),
    quartered or sliced
2/3 cup dry white wine
5 cups hot chicken stock
3 tablespoons freshly grated
    Parmesan cheese
Salt and pepper to taste

To Garnish:
Chopped parsley
Parsley sprigs

To Serve:
Tomato salad

Place cepes in a small bowl; cover with warm water and let soak 20 minutes. Drain; rinse well to remove any grit. Chop coarsely.

In a saucepan, melt 1/4 cup butter. Add onion and sauté about 5 minutes or until golden. Stir in rice and fresh mushrooms and cook 2 to 3 minutes or until rice is translucent. Add wine and cepes; cook about 3 minutes or until wine is absorbed. Reduce heat, add 2-1/2 cups stock, cover and simmer about 10 minutes or until almost all stock has been absorbed.

Add 1-1/4 cups more stock to pan and continue to cook, checking periodically and adding more stock as needed, until rice is tender and all liquid has been absorbed. Total cooking time will range from 20 to 30 minutes. Stir in remaining 1/4 cup butter, cheese, salt and pepper.

Transfer risotto to warmed serving plates and garnish with chopped parsley and parsley sprigs. Serve with a tomato salad.

Makes 4 to 6 servings.

Note: Dried cepes give this dish an incomparable flavor; for visual appeal and texture, I like to add fresh mushrooms. Use sliced fresh cepes (if available) or quartered button mushrooms. Avoid brown mushrooms, since they tend to color the risotto gray.

Stuffed Mushrooms

3/4 cup Brazil nuts, toasted, very
   finely chopped
1 teaspoon olive oil
1/2 teaspoon salt
8 large (2- to 3-inch-diameter)
   mushrooms (see *Note*)
1/4 cup butter
2 onions, chopped
1 garlic clove, pressed or minced
1 tablespoon chopped fresh thyme
1 cup fresh white bread crumbs
Juice and grated peel of 1 lemon
1/2 cup grated sharp Cheddar or
   Gruyère cheese
Salt and pepper to taste

To Garnish:
Thyme sprigs
Lemon slices

To Serve:
Tossed green salad

Preheat broiler. Combine nuts, oil and 1/2 teaspoon salt; set aside.

Remove stems from mushrooms; chop stems and set aside.

In a large frying pan, melt butter; brush half the melted butter over mushroom caps. Place caps, stemmed side down, on a baking sheet.

Reheat remaining butter. Add onions and garlic; sauté 5 to 7 minutes or until onions are softened and golden. Add chopped thyme and chopped mushroom stems; cook 1 minute. Transfer to a bowl; stir in nuts, crumbs, lemon juice, lemon peel, cheese, salt and pepper. Set aside.

Broil buttered mushroom caps 3 minutes, turning once. Spoon stuffing evenly onto stemmed sides of caps; broil 4 to 6 minutes or until stuffing is very hot.

Serve immediately, garnished with thyme sprigs and lemon slices. Accompany with a green salad.

Makes 4 servings.

Note: Freshly picked wild meadow mushrooms are preferable to cultivated mushrooms in this dish.

Mushroom & Herb Roulade

**6 ounces mushrooms (see *Note*),
    finely chopped**
1 bay leaf
1-1/4 cups milk
1/4 cup butter
1/2 cup all-purpose flour
1/2 cup grated Swiss cheese
1/2 teaspoon Dijon-style mustard
4 eggs, separated
1 hard-cooked egg yolk, sieved
**2 tablespoons finely snipped chives,
    dill or parsley**
2 tablespoons whipping cream
Salt and pepper to taste

To Garnish:
Herb sprigs

Preheat oven to 400F (205C). Grease a 13" x 9" x 2" baking pan and line with parchment paper.

In a saucepan, combine mushrooms, bay leaf and half the milk. Bring to a boil; reduce heat and simmer 2 minutes. Remove from heat, cover and let stand 10 to 15 minutes. Strain, reserving mushrooms and liquid; discard bay leaf. Add remaining milk to liquid.

In same pan, melt butter; stir in flour and cook, stirring, 1 minute. Remove from heat; gradually stir in milk mixture. Cook, stirring, until thick and smooth. Stir in cheese and mustard. Transfer a third of the sauce to a bowl; keep warm.

To sauce left in pan, add reserved mushrooms; then add raw egg yolks, one at a time, stirring well. In a clean bowl, whisk egg whites until stiff; fold into mushroom mixture. Pour into prepared pan and bake 12 to 15 minutes or until golden and firm. Turn out onto waxed paper; peel off parchment paper.

Stir sieved egg yolk, herbs, cream, salt and pepper into reserved sauce. Spread over roulade; roll up from a short side. Slice, garnish with herb sprigs and serve at once, with a green salad.

Makes 4 servings.

Note: Use parasols or horse mushrooms if you can get them; otherwise, use cultivated varieties.

Hot Mushroom Soufflé

1/4 cup butter
8 ounces mushrooms (see *Note*),
   chopped
1/2 cup all-purpose flour
1-1/4 cups milk
3/4 cup grated Swiss cheese
Salt and pepper to taste
3 eggs, separated

To Garnish:
Parsley sprigs

Preheat oven to 425F (220C). Grease a deep 4-cup soufflé dish.

In a saucepan, melt butter. Add mushrooms and sauté about 5 minutes or until quite tender. If mushrooms exude a lot of liquid, remove them with a slotted spoon and boil liquid over high heat until reduced, then return mushrooms to pan.

Add flour to pan; cook, stirring constantly, 2 minutes. Remove from heat and gradually stir in milk. Return to heat and cook, stirring, until smooth and very thick; stir in cheese. Remove from heat again and season with salt and pepper. Let cool slightly; then beat in egg yolks, one at a time.

In a clean bowl, whisk egg whites until stiff; carefully fold into mushroom mixture. Pour into prepared soufflé dish and bake on center rack of oven 40 to 45 minutes or until well risen, browned, and firm on top. Serve immediately, garnished with parsley sprigs.

Makes 4 servings.

Note: Horn of plenty, chanterelles, horse mushrooms and the anise-flavored *Agaricus silvicola* are particularly suitable for this recipe.

Baked Stuffed Avocados

1 tablespoon butter
8 ounces mushrooms (see *Note*),
    chopped
1 tablespoon plus 1 teaspoon
    all-purpose flour
Pinch of dry mustard
2/3 cup milk
1/2 cup grated sharp Cheddar cheese
Salt and pepper to taste
4 small ripe avocados
1 egg, separated

Herbed Crumbs:
2 tablespoons butter
1 cup fresh white bread crumbs
Pinch of celery salt
2 teaspoons chopped fresh dill
2 teaspoons chopped parsley

To Garnish:
Assorted salad greens

Preheat oven to 400F (205C). Prepare Herbed Crumbs: In a small saucepan, melt butter; add crumbs and cook, stirring, 3 to 4 minutes or until just beginning to color. Remove from heat and stir in celery salt, dill and parsley; set aside.

In another saucepan, melt 1 table-spoon butter. Add mushrooms and sauté 2 to 3 minutes or until tender. Stir in flour and mustard; cook, stirring, 1 minute. Remove from heat and gradually stir in milk; return to heat and cook, stirring, until thick and smooth. Stir in cheese. Remove from heat; add salt and pepper.

Halve and pit avocados. Scoop out some of flesh, leaving a 1/2-inch-thick shell; coarsely chop flesh and add to sauce. Stir in egg yolk.

In a clean bowl, whisk egg white until stiff; fold into sauce. Divide mixture among avocado shells; place shells on a baking sheet. Sprinkle with Herbed Crumbs and bake about 25 minutes or until hot. Serve immediately, garnished with salad greens.

Makes 4 servings.

Note: Parasols, chanterelles, hedgehog fungus and shaggy manes are ideal for this recipe. If you use cultivated mushrooms, try brown rather than white varieties.

Brie & Mushroom Tart

Tart Pastry:
2 cups all-purpose flour
Pinch of salt
1/2 cup butter

Spinach-Brie Filling:
2 tablespoons butter
6 ounces mushrooms (see *Note*),
  sliced
8 ounces spinach leaves, rinsed,
  shredded
10 ounces Brie cheese, rind removed,
  diced
3 eggs
2/3 cup half-and-half
Salt and pepper to taste

To Serve:
Tossed green salad

To make pastry, sift flour and salt into a large bowl; cut in butter until mixture resembles fine crumbs. Add about 3 tablespoons water and mix to form a firm dough.

Roll out dough on a lightly floured surface and use to line a greased 8-inch tart pan; prick shell lightly and chill 15 minutes. Meanwhile, preheat oven to 400F (205C). Bake tart shell blind 10 to 12 minutes. Remove from oven; reduce temperature to 350F (175C).

Prepare filling: In a large frying pan, melt butter. Add mushrooms and sauté 1 to 2 minutes. Add spinach and cook 1 to 2 minutes or until wilted. Turn out into a sieve and press out excess liquid; transfer spinach mixture to pastry shell. Tuck cheese cubes into spinach mixture. Beat together eggs, half-and-half, salt and pepper; pour over spinach mixture. Bake 30 to 35 minutes or until filling is set.

Serve warm or cold, with salad. (Tart tastes best warm.)

Makes 6 servings.

Note: Use button mushrooms or wild varieties such as chanterelles, horse mushrooms or parasols; or use a mixture.

Spinach & Feta in Filo

1/4 cup butter
1 shallot, finely chopped
1 garlic clove, pressed or minced
12 ounces mushrooms (see *Note*),
    finely chopped
1/3 cup white wine
1 tablespoon plus 1 teaspoon tomato
    paste
1 tablespoon chopped fresh dill
Salt and pepper to taste
8 ounces spinach leaves, rinsed,
    chopped
8 ounces feta cheese, crumbled
1 egg yolk
Pinch of freshly grated nutmeg
8 sheets filo pastry

Preheat oven to 400F (205C). In a saucepan, melt 2 tablespoons butter. Add shallot and garlic and sauté 2 minutes. Add mushrooms, wine and tomato paste; cook over high heat, stirring frequently, 8 to 10 minutes or until all liquid has evaporated. Stir in dill, salt and pepper; set aside.

In another pan, cook spinach, with just the water that clings to the leaves, 2 to 3 minutes or until wilted. Turn out into a sieve and press out excess liquid; transfer spinach to a bowl. Add cheese, egg yolk and nutmeg; mix well.

Melt remaining 2 tablespoons butter. Brush a baking sheet with some of butter; place a sheet of pastry on baking sheet. Brush lightly with butter, cover with a second sheet of pastry and brush again. Spread mushroom mixture over pastry, leaving a 1-inch border; cover with spinach mixture. Dampen pastry edges; cover fillings with remaining 6 pastry sheets, brushing each sheet with butter. Seal edges. With a sharp knife, score a deep cross on surface. Bake about 25 minutes or until pastry is crisp and golden brown.

Serve hot or warm, quartered or in slices, with salad and new potatoes.

Makes 4 servings.

Note: Use meadow mushrooms, horse mushrooms or parasols; or use cultivated button mushrooms.

Scrambled Eggs with Truffles

6 eggs
1/2 ounce fresh or canned white
   truffles
Salt and pepper to taste
6 to 8 fresh asparagus spears
2 slices whole wheat bread
3 tablespoons butter
2 tablespoons whipping cream

To Garnish:
Chervil sprigs

In a bowl, beat eggs until blended. Cut truffles into very fine slivers; set half aside. Add remaining slivers to eggs; if using canned truffles, add any juices as well. Season with salt and pepper, cover and refrigerate at least 2 hours.

When you are almost ready to cook eggs, prepare asparagus. Break off tough stalk ends; using a potato peeler, thinly peel stalks. Cook in boiling water 4 to 5 minutes or until just tender; drain and keep hot.

Meanwhile, toast bread and spread with 1-1/2 tablespoons butter; keep hot.

In a deep nonstick frying pan, melt remaining 1-1/2 tablespoons butter over low heat. Add egg-truffle mixture and cook, stirring lightly with a wooden spoon, about 3 minutes or until eggs begin to thicken and set. Take care not to overcook—eggs should be creamy and soft. Stir in cream.

Arrange asparagus and hot toast on 2 warmed serving plates and divide scrambled eggs between them. Sprinkle with reserved truffle slivers. Garnish with chervil sprigs and serve immediately.

Makes 2 servings.

Omelet with Chanterelles

5 eggs
2 tablespoons chopped fresh herbs,
    such as parsley, chervil, chives,
    tarragon or dill
3 tablespoons water
2 tablespoons butter

Chanterelle Filling:
1 tablespoon butter
1 garlic clove, halved
8 ounces chanterelles
2 tablespoons whipping cream
Salt and pepper to taste

To Garnish:
Herb sprigs

Prepare filling: In a small frying pan, melt butter. Add garlic and cook gently 1 minute, being careful not to burn butter. Discard garlic. Add chanterelles to pan and cook over high heat 5 minutes. If mushrooms exude a lot of liquid, remove them with a slotted spoon and boil liquid over high heat until reduced, then return mushrooms to pan.

Remove from heat, stir in cream and season with salt and pepper. Keep warm while preparing omelet.

In a bowl, beat together eggs, chopped herbs and 3 tablespoons water. In a nonstick frying pan, melt butter; pour in eggs. When edges begin to set, gently lift them with a spatula and push toward center, allowing uncooked eggs to flow underneath. Continue to cook in this way until omelet is almost set.

Spoon Chanterelle Filling over half the omelet, then fold other half over to enclose. Cut omelet in half and slide onto warmed serving plates. Garnish with herb sprigs and serve immediately.

Makes 2 servings.

Note: Eggs for omelets are exquisite when scented by truffles. Simply store a truffle with the eggs; its aroma will penetrate the shells.

Tagliatelle with Mushrooms

1/2 ounce dried cepes (porcini)
2 tablespoons butter
8 ounces small button mushrooms,
  quartered
1 garlic clove, pressed or minced
2/3 cup dry white wine
2 eggs
1/2 cup whipping cream
1 pound fresh tagliatelle
3 to 4 tablespoons freshly grated
  Parmesan cheese
Salt and pepper to taste

To Garnish:
Chopped parsley
Parsley sprigs

Place cepes in a small bowl; cover with warm water and let soak 20 minutes. Drain; rinse well to remove any grit. Chop coarsely.

In a saucepan, melt butter. Add button mushrooms and garlic; sauté 2 minutes. Add cepes and wine; cook over high heat 2 to 3 minutes or until mushrooms are tender and liquid is reduced by half. Keep hot.

Beat eggs and cream together; set aside.

Cook tagliatelle in plenty of boiling salted water about 3 minutes or just until *al dente;* drain well and return to pan. Add eggs and hot mushroom mixture to pasta and toss until eggs become creamy and start to set. Add cheese, salt and pepper and toss again. Serve immediately, garnished with chopped parsley and parsley sprigs.

Makes 4 servings.

Note: If you are fortunate enough to find fresh cepes or other boletes, use them in place of both dried cepes and fresh button mushrooms. You will need about 8 ounces; chop before using.

Pasta, Beans & Mushrooms

1 pound unshelled fava beans, shelled
8 ounces dried rigatoni
2 tablespoons butter or virgin olive
  oil
Salt to taste

Mushroom Sauce:
2 tablespoons virgin olive oil
2 shallots, finely chopped
1 garlic clove, pressed or minced
2 to 3 tablespoons chopped sun-dried
  tomatoes (see *Note*, page 75)
12 ounces wild mushrooms (see *Note*),
  chopped if large
2 tablespoons finely chopped parsley
1 cup tomato puree
1/4 cup whipping cream
Salt and pepper to taste

To Garnish:
Freshly grated Parmesan cheese
Parsley sprigs

Prepare Mushroom Sauce: In a
saucepan, heat oil. Add shallots and
garlic and sauté 3 to 4 minutes or
until softened and just beginning to
color. Stir in sun-dried tomatoes,

mushrooms and parsley; cook 2 minutes, stirring constantly. Stir in tomato puree and simmer 5 to 6 minutes
or until mushrooms are tender. Stir
in cream, salt and pepper; heat gently. Set aside.

Cook beans in boiling water 30
minutes or until tender. In another
pan, cook rigatoni in plenty of boiling
salted water 8 to 10 minutes or until *al
dente*. Drain pasta and beans thoroughly and put in a large warmed
serving bowl. Toss together with butter or oil and a little salt, if desired.

Gently reheat sauce; do not boil.
Add half the sauce to bowl; toss lightly to mix, then spoon onto warmed
plates and offer remaining sauce to
spoon on top. Garnish with cheese
and parsley sprigs.

Makes 4 servings.

Note: Use meadow or parasol mushrooms, morels, or cepes (porcini) or
other large boletes.

Mushroom Stroganoff

2 tablespoons butter
2 tablespoons virgin olive oil
2 onions, thinly sliced
1 or 2 garlic cloves, pressed or
    minced
1-1/2 pounds cepes (porcini) or large
    meadow mushrooms, sliced
1/4 cup brandy
1-1/4 cups dairy sour cream
Salt and pepper to taste
Paprika

To Garnish:
Whole chives

To Serve:
Saffron rice or crusty bread

In a large frying pan, melt butter in oil. Add onions and cook 5 to 7 minutes or until softened and golden. Add garlic and mushrooms; cook 2 minutes, stirring constantly.

Stir in brandy and cook, stirring frequently, 3 to 4 minutes or until mushrooms are tender. Reduce heat and stir in sour cream; heat gently, being careful not to boil. Season with salt and pepper.

Sprinkle with paprika and garnish with chives. Serve with saffron rice or warm crusty bread.

Makes 4 servings.

Note: Large, meaty-textured mushrooms are essential for this dish. For appearance as well as taste, your best choices are cepes or other boletes. Meadow mushrooms have a good flavor, but they'll color the sauce gray. Cultivated mushrooms aren't as flavorful as wild ones, but can give good results with the help of dried cepes or suillus.

Mushroom & Broccoli Stir-Fry

4 ounces firm tofu, cut into 3/4-inch
   cubes
2 tablespoons soy sauce
4 dried shiitake mushrooms
2 to 3 tablespoons peanut or olive oil
8 ounces broccoli flowerets
1 garlic clove, pressed or minced
2 teaspoons shredded fresh
   gingerroot
1 red onion, cut into thin wedges
1 (4-oz.) jar straw mushrooms,
   drained
3 tablespoons dry sherry
1 teaspoon cornstarch
Generous pinch of sugar
Black or Szechuan pepper to taste

To Garnish:
1 teaspoon sesame seeds, toasted
Cilantro sprigs

To Serve:
Hot rice or thin noodles

In a small bowl, toss tofu with soy sauce; set aside. Place shiitake in a another small bowl, cover with 1/2 cup hot water and let soak 20 minutes. Drain, adding soaking liquid to tofu. Rinse shiitake well; trim off tough stems and slice caps.

In a wok, heat oil over high heat. Using a slotted spoon, remove tofu from marinade; reserve marinade. Add tofu to wok and stir-fry 1 minute or until browned; remove and set aside.

Add broccoli, garlic and ginger to wok and stir-fry 2 minutes. Add onion, shiitake and straw mushrooms and stir-fry 1 minute. Pour in tofu marinade and bring to a boil; simmer 2 minutes or until broccoli is tender-crisp. Stir in tofu.

Blend sherry, cornstarch and sugar; pour into wok and cook, stirring, until thickened. Season with pepper; transfer to a warmed serving dish.

Garnish with sesame seeds and cilantro sprigs. Serve immediately, with rice or noodles.

Makes 4 servings.

Note: Dried wood ears may also be added for their pretty, frilly appearance and chewy texture. Reconstitute them with the shiitake.

Wild Rice Salad

1-1/4 cups quick-cooking cracked
    wheat
3 tablespoons wild rice, rinsed,
    drained
4 ounces Gruyère cheese, diced
4 ounces Pipo Crème or other creamy
    blue cheese, diced
10 cherry tomatoes, halved
2 green onions, shredded
1/4 cup virgin olive oil
8 ounces mixed chanterelles and
    oyster mushrooms, quartered if
    large
1 garlic clove, chopped
2 tablespoons white wine vinegar
2 tablespoons chopped fresh dill
Pinch of sugar (optional)
Salt and pepper to taste

To Garnish:
Dill sprigs

Place cracked wheat in a bowl; pour
1-1/4 cups cold water over it. Let
stand about 1 hour or until wheat is
tender, but still chewy.

Meanwhile, in a small saucepan,
combine wild rice and 1 cup water.
Bring to a boil; then reduce heat,
cover and simmer about 45 minutes
or until rice is tender. Drain; let cool.

Drain any unabsorbed water from
wheat. In a large salad bowl, combine
wheat, rice, cheeses, tomatoes and
green onions.

In a large frying pan, heat oil. Add
mushrooms and sauté 4 minutes. Remove from heat and add garlic, vinegar, chopped dill and sugar, if desired; season with salt and pepper.
Add to salad and toss lightly to mix.
Serve immediately; or chill and serve
cold. Garnish with dill sprigs before
serving.

Makes 4 to 6 servings.

Saffron Rice Pilaf

Generous pinch of saffron threads
1 tablespoon sesame or olive oil
1/2 cup slivered almonds
3 tablespoons butter
1 onion, chopped
1 garlic clove, pressed or minced
1 large carrot, cut into matchstick
    pieces
1 celery stalk, cut into matchstick
    pieces
1-1/2 cups brown rice
10 ounces mixed chanterelles and
    oyster mushrooms, quartered if
    large
1/3 cup golden raisins
2-1/2 cups mushroom or vegetable
    stock
4 ounces snow peas, halved
1 to 2 tablespoons chopped fresh
    cilantro
Salt and pepper to taste

To Garnish:
Fresh cilantro leaves
Marigold petals (optional)

Place saffron in a small bowl, cover with 1/4 cup boiling water and let soak 20 minutes.

In a saucepan, heat oil. Add almonds and cook, stirring, 1 to 2 minutes or until browned. Transfer to a plate and set aside.

Melt butter in pan over medium heat. Add onion, garlic, carrot and celery; cook 3 to 4 minutes or until vegetables just begin to soften. Add rice and cook 4 minutes, stirring often. Stir in mushrooms, raisins, stock and saffron with its soaking liquid. Bring to a boil; reduce heat, cover and cook 25 to 30 minutes or until rice is tender.

Stir in snow peas, chopped cilantro, toasted almonds, salt and pepper.

To serve, garnish pilaf with cilantro leaves and sprinkle with marigold petals, if desired.

Makes 4 servings.

Italian Mushroom Salad

1 small head red oak leaf or lollo
   rosso lettuce
1/2 red onion, thinly sliced
8 ounces button mushrooms, thinly
   sliced
2 ounces Parmesan cheese
2 teaspoons finely chopped parsley

Lemon Dressing:
5 tablespoons virgin olive oil
Finely grated peel of 1/2 lemon
Juice of 1 small lemon
1/4 teaspoon coarse-grained mustard
Pinch of sugar
Salt and pepper to taste

Prepare Lemon Dressing: Stir all dressing ingredients together in a small bowl, or combine in a screw-top jar and shake well. Set aside.

Tear lettuce into bite-size pieces; arrange lettuce and onion on individual serving plates. Set aside.

Place mushrooms in a large bowl. Pour dressing over them and toss well to coat. Pare cheese into wafer-thin slices and add to mushrooms, tossing lightly to mix.

Arrange mushroom mixture atop lettuce, sprinkle with parsley and serve immediately.

Makes 4 servings.

Note: The mushrooms may be left to marinate in the dressing up to 2 hours, but it's best to add the Parmesan to the salad just before serving.

Pepper & Mushroom Salad

1 each green, red and yellow bell
   pepper, halved, seeded
2 tablespoons virgin olive oil
1 shallot, finely chopped
1 garlic clove, pressed or minced
6 ounces wild mushrooms (see *Note*),
   cut into large pieces
1 to 2 tablespoons red wine vinegar
1/2 teaspoon Dijon-style mustard
Salt and pepper to taste
2 ounces feta cheese, crumbled

To Garnish:
Parsley sprigs

Preheat broiler. To prepare peppers, place on a baking sheet, cut sides up; broil 3 minutes. Turn and broil 5 more minutes or until skin blisters and blackens. Peel off skin, cut peppers into 3/4-inch pieces and arrange on individual serving plates. Set aside.

   In a frying pan, heat oil. Add shal-lot and garlic and sauté 2 minutes or until softened. Add mushrooms and cook 2 minutes. Using a slotted spoon, arrange mushroom mixture over peppers.

   Add vinegar and mustard to pan juices; boil over high heat until reduced to 1 to 2 tablespoons. Season with salt and pepper, then drizzle over peppers and mushrooms. Let cool, then chill at least 30 minutes. Before serving, adjust seasoning; sprinkle with cheese and garnish with parsley sprigs.

Makes 4 servings.

Note: A mixture of wild mushrooms is ideal for this salad. If wild types are unavailable, try a combination of cultivated oyster mushrooms and sliced button mushrooms.

Asparagus & Prawn Salad

4 ounces dried penne or other pasta
  shapes
1 pound asparagus
4 ounces button mushrooms, sliced
6 ounces shelled, deveined cooked
  prawns
1 green onion, shredded
1/2 cup dairy sour cream
Salt and pepper to taste

To Garnish:
Lemon slices
2 to 4 cooked prawns in shell

Following package directions, cook
pasta in boiling salted water until *al
dente*. Rinse under cold running water
and drain thoroughly. Transfer to a
large bowl.

Break off tough stalk ends from
asparagus. Cut off tips; set aside. Divide asparagus stalks into 2 equal por-
tions; thinly slice one portion and set
the other aside. Cook asparagus tips
and sliced stalks in boiling water 1
minute. Rinse in cold water; add to
pasta with mushrooms, shelled
prawns and green onion.

Cook remaining asparagus stalks in
boiling salted water 6 to 7 minutes or
until soft. Drain well; transfer to a
food processor or blender, add sour
cream and puree until smooth. Add
to pasta salad and toss to mix; season
with salt and pepper.

Transfer to individual serving
plates and garnish with lemon slices
and unshelled prawns.

Makes 2 to 4 servings.

Mediterranean Lamb Salad

1 small (about 8-oz.) eggplant, diced
Salt
2 tablespoons virgin olive oil
1 small onion, thinly sliced
2 zucchini, thinly sliced
6 ounces wild mushrooms (see *Note*),
   sliced if large
4 tomatoes, peeled, seeded, quartered
12 pitted ripe olives
12 ounces rare roast lamb, cut into
   strips

Herb Dressing:
3 tablespoons virgin olive oil
2 tablespoons red wine vinegar
2 teaspoons Dijon-style mustard
1 tablespoon chopped fresh rosemary
1 tablespoon chopped fresh thyme
Pinch of sugar
Salt and pepper to taste

Prepare Herb Dressing: Stir all dressing ingredients together in a small bowl, or combine in a screw-top jar and shake well. Set aside.

Rinse eggplant with cold water and place in a colander. Sprinkle with salt and let stand 20 minutes. Rinse to remove salt; pat dry with a clean dishtowel.

In a large frying pan or wok, heat 1 tablespoon oil. Add onion and sauté 3 minutes. Add eggplant and zucchini and cook 4 to 5 minutes or until softened. Transfer to a large salad bowl.

In same pan, heat remaining 1 tablespoon oil. Add mushrooms and sauté 2 to 3 minutes or until tender; pour off any excess liquid. Add mushrooms, tomatoes, olives and lamb to salad bowl. Pour dressing over salad and toss well to mix. Chill at least 20 minutes before serving.

Makes 4 to 6 servings.

Note: Use a mixture of wild mushrooms, such as cauliflower fungus, chanterelles, morels, hedgehog fungus and, if you are certain of their identity, any of the russulas.

Cherry Tomato & Bacon Salad

1 pound cherry tomatoes
5 ounces brown or chestnut
   mushrooms, sliced
2 tablespoons olive oil
6 bacon slices, diced
1 tablespoon white wine vinegar
3 tablespoons dairy sour cream
Salt and pepper to taste

To Garnish:
4 pitted ripe olives, chopped
1 tablespoon snipped chives
Whole chives

Plunge half the tomatoes into a bowl of boiling water; leave for 30 seconds, then drain and peel. Repeat with remaining tomatoes. Transfer to a salad bowl and add mushrooms.

In a large frying pan, heat oil. Add bacon and cook about 3 minutes or until crisp; lift from pan with a slotted spoon and transfer to salad bowl. Deglaze pan with vinegar, stirring to scrape up browned bits; add pan drippings to salad with sour cream, salt and pepper. Toss lightly to mix and chill at least 15 minutes.

To serve, garnish with olives, snipped chives and whole chives.

Makes 4 to 6 servings.

Note: I like the earthy flavor of brown or chestnut mushrooms in this salad. If they aren't available, you can substitute button mushrooms.

Salade Tiède

**12 ounces mixed salad greens, such
as mâche (lamb's lettuce),
radicchio, red oak leaf, romaine,
chicory and escarole**
1 red onion, thinly sliced
1/4 cup hazelnut or virgin olive oil
3 tablespoons pine nuts
4 bacon slices, diced
**1 small garlic clove, pressed or
minced**
**5 ounces wild mushrooms (see *Note*),
sliced**
2 tablespoons raspberry vinegar
Pinch of sugar
**1/2 teaspoon Dijon-style or tarragon
mustard**
Salt and pepper to taste

Tear salad greens into bite-size
pieces; place greens and onion in a
salad bowl.

In a large frying pan, heat 1 tablespoon oil. Add pine nuts and cook
about 2 minutes or until golden
brown, stirring constantly. Drain on
paper towels and set aside.

Add bacon to pan and cook about 3
minutes or until crisp. Transfer to a
plate; set aside.

Heat 1 more tablespoon oil in pan;
add garlic and mushrooms. Cook 2 to
3 minutes or until tender. If mushrooms exude a lot of liquid, remove
them with a slotted spoon to a plate
and boil liquid over high heat until
reduced to 2 tablespoons, then return
mushrooms to pan.

Return bacon to pan and add
remaining 2 tablespoons oil, vinegar,
sugar, mustard, salt and pepper.
Heat for a few seconds, stirring; then
add to salad greens and toss lightly.
Sprinkle with toasted pine nuts and
serve immediately.

Makes 4 servings.

Note: Use one or more kinds of small,
pretty mushrooms such as chanterelles, fairy rings, or pieces of cauliflower fungus.

Potato & Mushroom Salad

2 pounds small thin-skinned potatoes,
   peeled, boiled just until tender
   throughout, drained
1 small red onion
2 tablespoons virgin olive oil
6 ounces cepes (porcini, see *Note*),
   coarsely chopped
1 tablespoon white wine vinegar
2 tablespoons chopped fresh dill
Salt and pepper to taste
1/3 cup whipping cream
1 tablespoon coarse-grained mustard

To Garnish:
Dill sprigs

Let hot boiled potatoes stand until cool enough to handle, then thickly slice into a large bowl.

Cut onion into wedges and separate each wedge into layers.

In a large frying pan, heat oil. Add onion and cepes; sauté 2 minutes.

Add vinegar and cook 1 to 2 more minutes or until mushrooms are tender; remove onions and mushrooms from pan with a slotted spoon and add to potatoes.

If mushrooms have exuded a lot of liquid, boil liquid over high heat until reduced to 2 tablespoons. Add to salad with chopped dill, salt and pepper; toss lightly to mix, then transfer to a serving plate.

Mix cream and mustard and drizzle over salad. Serve warm or cold, garnished with dill sprigs.

Makes 6 servings.

Note: If cepes are not available, you can substitute meadow or button mushrooms, though the flavor won't be as good.

Orange & Tarragon Mushrooms

3 tablespoons olive oil
1 shallot, finely chopped
1 pound small button mushrooms
1 (1-inch) piece fresh gingerroot,
  grated
1 garlic clove, pressed or minced
1 tablespoon sesame oil
2 tablespoons balsamic vinegar
1 teaspoon grated orange peel
Juice of 1 orange
Salt and pepper to taste

To Serve:
2 celery stalks, cut into thin julienne
  strips
1 tablespoon chopped fresh tarragon

In a large frying pan, heat 2 tablespoons olive oil. Add shallot and sauté 3 to 4 minutes or until lightly browned. Add mushrooms, ginger and garlic; cook, stirring frequently, 4 to 5 minutes or until mushrooms are tender. Using a slotted spoon, transfer vegetables to a bowl.

To liquid in pan, add sesame oil, vinegar, orange peel and orange juice. Boil over high heat until reduced by half. Pour mixture over mushrooms and let cool. Add salt and pepper; then chill at least 30 minutes.

To serve, arrange mushrooms on a serving plate. Surround with celery and sprinkle with tarragon.

Makes 4 servings.

Scented Garlic Mushrooms

12 ounces button mushrooms
2/3 cup white wine
1 bay leaf
Seeds of 2 white cardamom pods
2 garlic cloves, pressed or minced
3 to 4 tablespoons dairy sour cream
Salt and pepper to taste

To Serve:
Young spinach leaves

In a large saucepan, combine mushrooms, wine, bay leaf, cardamom seeds and half the garlic. Bring to a boil, then reduce heat and simmer, uncovered, about 10 minutes or until mushrooms are tender and liquid is reduced by about half.

Let cool slightly, then discard bay leaf and stir in sour cream and remaining garlic. Season with salt and pepper. Chill at least 30 minutes before serving.

Serve mushrooms with young spinach leaves.

Makes 4 servings.

Baked Mushrooms in Madeira

1/4 cup butter, room temperature
1 pound large meadow mushrooms,
  thickly sliced
1 large garlic clove, thinly slivered
Salt and pepper to taste
2/3 cup Madeira

To Garnish:
1 tablespoon chopped parsley

Preheat oven to 375F (190C). Spread butter over bottom of a large, shallow baking dish. Arrange mushrooms in a single layer in dish and dot evenly with garlic slivers. Season with salt and pepper; pour Madeira over top.

  Bake, covered, 25 to 30 minutes or until mushrooms are very tender. Serve hot or cold, sprinkled with parsley.

Makes 4 servings.

Note: Fresh cepes (porcini) are delicious cooked this way. Cultivated brown mushrooms can also be used, but their flavor is not as rich as that of the wild variety.

Mushrooms in Beer Batter

Vegetable oil for deep-frying
1 pound mushrooms (see *Note*)
Sea salt
Lemon wedges

Beer Batter:
1 cup all-purpose flour
1/2 teaspoon salt
2 eggs, separated
2/3 cup beer
2 tablespoons sunflower oil

To Garnish:
Chervil sprigs

To prepare batter, sift flour and salt into a bowl; add egg yolks, then gradually whisk in beer to form a smooth batter. Stir in oil. In a clean bowl, whisk egg whites until they hold soft peaks, then fold into batter.

In a deep, heavy saucepan, heat about 3 or 4 inches of oil to 350F (175C). Dip mushrooms into batter to coat; lower into hot oil, a few at a time (do not crowd pan). Cook 2 to 3 minutes or until crisp and golden. Drain on paper towels and keep hot while cooking remaining mushrooms.

Serve hot, sprinkled with sea salt and fresh lemon juice and garnished with chervil sprigs.

Makes 4 servings.

Note: Choose a mixture of wild and cultivated mushrooms for this accompaniment, aiming for a pretty assortment of shapes. Try button mushrooms, oyster mushrooms, morels, shiitake, parasols, chanterelles and sliced small puffballs.

Eggplant Gâteaux

1 large eggplant, thinly sliced
3 tablespoons garlic salt
1-1/4 cups tomato puree
3 to 4 tablespoons virgin olive oil
12 ounces mushrooms (see *Note*),
   finely chopped
1 garlic clove, pressed or minced
1 tablespoon chopped fresh basil
2 tablespoons brandy
Salt and pepper to taste
1/4 cup plain yogurt

To Garnish:
Flat-leaf parsley and basil sprigs

Preheat oven to 350F (175C). Rinse eggplant with cold water; place in a colander. Sprinkle with garlic salt and let stand 20 minutes. Rinse to remove salt; pat dry with a clean dishtowel.

Meanwhile, in a saucepan, boil tomato puree until reduced by half; set aside.

In a large frying pan, heat 2 to 3 tablespoons oil and sauté eggplant (in batches, if necessary) 4 to 5 minutes or just until softened. Use slices to line 4 (4-inch) ramekins, reserving enough slices to cover filling.

In same pan, heat remaining 1 to 2 tablespoons oil over high heat; add mushrooms, garlic and basil. Cook 10 to 12 minutes or until all liquid has evaporated. Add brandy and cook 2 minutes. Season with salt and pepper and divide among prepared ramekins. Top with yogurt, then with tomato puree. Overlap eggplant slices to enclose filling; press down lightly. Cover ramekins with foil; bake 20 minutes.

Turn out of ramekins and garnish with parsley and basil sprigs.

Makes 4 servings.

Note: Use meadow or horse mushrooms or cepes (porcini), or a mixture of wild varieties, rather than cultivated kinds.

Serve these gâteaux as a stylish accompaniment to lamb or chicken. They can be assembled in advance, then baked just before serving.

There are about seventy amateur societies in North America for people who are interested in mushrooms. One, the North American Mycological Association (affectionately known to its members as NAMA), is national in scope. The other groups are regional; most of them publish newsletters, organize mushroom-hunting forays and give beginners a chance to learn identification from more experienced enthusiasts. Membership typically runs from ten to twenty dollars per year. NAMA can send you information about societies in your area; write to the address below, being sure to enclose a stamped, self-addressed envelope with your letter.

Executive Secretary
North American Mycological Association
3556 Oakwood
Ann Arbor, MI 48104-5213

It would be impossible to list all the many books in print about mushrooms; visit a library or bookstore to find out what is available. Because mushrooms' seasons and habitats vary greatly, a regional guide is the best index to the mushrooms you may find locally. Even when a regional guide exists, though, you may want to complement it with a more comprehensive manual; a few recent ones are listed below. All of them explain important procedures—such as taking a spore print—and define the vocabulary used for describing mushrooms.

A beginner trying to find out more about any of the mushrooms in this book will obviously have trouble locating a description if another book gives the mushroom a different name. In fact, novices are likely to be surprised at how little agreement there is over names, sometimes even the Latin (botanical) ones. Nevertheless, the Latin names generally vary less than the so-called "common" names. It is not practical to list all the possibilities, but when you are lost, follow this useful clue: the second Latin name (the species epithet) is less likely to vary than the first (the name of the genus), though you will notice variation in the species-name suffix, depending on the grammatical gender of the generic name. For example, *Lepiota procera*, the parasol mushroom, is called *Leucoagaricus procerus* by Smith and *Leucocoprinus procera* by McKnight. (Incidentally, McKnight gets the gender-agreement wrong on this one—he should use *procerus*.) Roughly speaking, this situation reflects general agreement that there is a distinct kind of mushroom (a species) but disagreement over how to classify it (what genus to put it in). Of course, the same species name may be used in two different genera for genuinely different mushrooms, so read the descriptions carefully to see if the two names seem to refer to the same thing.

Arora, David. *Mushrooms Demystified*. 2nd Edition. Berkeley, California: Ten Speed Press, 1986.
An excellent, if weighty, book describing more species than any other field guide and leavened with a fine sense of humor. It has outgrown its origins in the San Francisco area, though the emphasis remains on California and "our area" still means "around Santa Cruz."

Lincoff, Gary H. *The Audubon Society Field Guide to North American Mushrooms* New York: Knopf, 1981.
Contains 756 color photographs, more than any other field guide, and makes a reasonable attempt to cover all of North America. Lincoff uses visual characters to help identify mushrooms, which is easier for beginners but—from the expert's point of view—less precise than the dichotomous keys, the systematic technique used in other books. Purists are offended by his coinage of a multitude of new "common" names. The book combines a pleasing layout, good coverage and reasonable portability.

McKnight, Kent H., and Vera B. McKnight. *A Field Guide to Mushrooms*. In *The Peterson Field Guide Series*. Boston: Houghton Mifflin, 1987.
Some people consider that a drawing or painting can bring out the distinctive character of a mushroom better than a photograph. This book contains about 500 excellent illustrations by Vera McKnight, about 460 of them in color. Descriptions are careful and thorough, but coverage is less than that provided by Lincoff and presentation is crowded.

Miller, Orson. *Mushrooms of North America*. New York: Dutton, 1972.
Orson Miller is an academic mycologist whose book is intermediate in coverage and somewhat technical. No common names are used, even well-established ones like "cepes."

Smith, Alexander H., and Nancy Smith Weber. *The Mushroom Hunter's Field Guide: Revised and Enlarged*. Ann Arbor: University of Michigan Press, 1963.
Alexander Smith was for many years a doyen of professional mycologists in North America, and this book was a landmark among field guides when it was first published in 1958. Descriptions are good, but the coverage is much less extensive than that in the other books listed. Smith followed this book with two comparably sized regional guides from the same press: *Western Mushrooms* (1975) and, with Nancy Smith Weber, *Southern Mushrooms* (1985).